Make: FPGAs

*Turning Software into Hardware with
Eight Fun and Easy DIY Projects*

David Romano

MAKER MEDIA
SAN FRANCISCO, CA

Make: FPGAs

by David Romano

Copyright © 2016 David Romano. All rights reserved.

Printed in the United States of America.

Published by Maker Media, Inc., 1160 Battery Street East, Suite 125, San Francisco, CA 94111.

Maker Media books may be purchased for educational, business, or sales promotional use. Online editions are also available for most titles (*http://safaribooksonline.com*). For more information, contact O'Reilly Media's institutional sales department: 800-998-9938 or *corporate@oreilly.com*.

Editor: Roger Stewart	**Interior Designer:** David Futato
Production Editor: Melanie Yarbrough	**Cover Designer:** Brian Jepson
Copyeditor: Rachel Head	**Cover Art:** Shawn Wallace
Proofreader: Christina Edwards	**Illustrator:** Rebecca Demarest
Indexer: Judy McConville	

February 2016:   First Edition

Revision History for the First Edition
2016-02-26:   First Release

See *http://oreilly.com/catalog/errata.csp?isbn=9781457187858* for release details.

978-1-457-18785-8

[LSI]

Table of Contents

Preface

Mention FPGAs to most people, and they will either give you a blank stare or think you are talking about some kind of golf league. To most of us Makers, the term conjures up thoughts of hardware creativity, exploration, and discovery, but many of you may have written it off as being way too complicated to even consider for your next project. This book is for you! It's all about learning what amazing, easy, and affordable projects you can construct with field programmable gate array (FPGA) technology. We will be doing this with hands-on experiments, in a fun and practical way.

This book is not a university textbook providing in-depth studies on hardware description languages (HDLs), HDL coding techniques, digital logic design theory, or validation methods. There are many very good resources both online and in textbook form that accomplish this goal. This will be more a learn-as-you-go experience. You can think of the book as a road map to a journey of design discovery, and I'll be your guide. But before we jump in, I want to give you a little background on the history of FPGAs.

Most of the technical community was first introduced to this exciting technology back in the '80s. I was a recent college graduate (RCG) with a degree in electrical engineering and had just been hired by a small telecommunications company that designed and manufactured modems and TI multiplexers. The company was developing a product that was, at the time, implemented using many 7400 series integrated circuits (ICs) and programmable array logic (PAL) devices. For many of you, this is probably like talking about funny animal paintings on cave walls. It wasn't the Stone Age of electronic design, but 7400 ICs and PALs are primitive compared to today's state-of-the-art systems-on-a-chip (SoCs). My new manager came to me one day with a data book in hand that he had just received from a company called Xilinx. He dropped the book on my desk and said, "Well, since you're the college kid, I want you to convert all of this logic design to an FPGA device," pointing to a large and very complex circuit board. Being a little naive and always eager to work on the latest cutting-edge technology, I said, "No problem, should be a piece of cake." It wasn't a piece of cake, but it was my introduction into the incredible world of field programmable array logic—sort of a baptism by fire.

A lot has happened in the past three decades in the world of digital design, and programmable logic devices (PLDs) are a big part of it. A PLD is an electronic component used to build reconfigurable digital circuits. Unlike logic gates, like those in the 7400 series ICs, which have fixed logic functions, a PLD has an undefined function at the time of manufacture. Before the PLD can be used in a circuit, it must be programmed (i.e., reconfigured). Before PLDs were invented, read-only memory (ROM) chips were used to create arbitrary combinational logic functions—talk about the Stone Age!

Today, there are many reasons a design team will consider FPGA technology in industry. For example, in many silicon IC design companies, FPGA-based platforms are used for what is called "shift left," where a new SoC device is mapped to FPGAs early in the design phase, in order to begin software integration long before the actual silicon device is manufactured. This is called "emulation" of the design. The big advantage here is that emulation runs orders of magnitude faster than simulation, so you can get some real-world hardware/software interactions very early in the validation phase (mostly on a functional level). The FPGA system typically operates at only a fraction of the silicon operating frequency, but the time saved in integration is tremendous.

Another example of where FPGAs are considered a viable solution in industry is where the design requires having multiple hardware personalities in the same footprint. For example, this was the case for a portable test and measurement instrument that I architected when I was a design engineer. By using an FPGA in the design, the customer was able to download different test instruments to the same hardware, essentially having multiple instruments in one hardware device.

The real question, then, for this book is: why would you, the do-it-yourself hobbyist or student, even consider experimenting with FPGAs? For students, it exposes you to contemporary digital logic design methods and practices in a fun, practical, affordable way. For the hobbyist, my goal is to show you how an affordable, off-the-shelf FPGA platform can be used in some very interesting and fun DIY projects. Many of you have had experience with Arduino or similar small microcontroller projects. With these types of projects, you usually breadboard up a small circuit, connect it to your Arduino, and write some code in the C programming language (which Arduino is based on) to perform the task at hand. Typically your breadboard can hold at best a few discrete components and one or two small ICs. Then you always need to go through the pain of wiring up the circuit and connecting it to your Arduino with a rat's nest of jumper wires. Instead, imagine having a breadboard the size of a basketball court or football field to play with and, best of all, no jumper wires. Imagine you can connect everything virtually. You don't even need to buy a separate microcontroller board; you can just drop different processors into your design as you choose. Now that's what I'm talking about! Welcome to the world of FPGAs.

FPGA History

Xilinx Inc. was founded in 1984, and as the result of numerous patents and technology breakthroughs, the company produced the first family of general-purpose, user-programmable logic devices based on an array architecture. It called this technology breakthrough the Logic Cell Array (LCA), and with this the Xilinx XC 2000 family of FPGAs was born.

You can think of an LCA as being made up of three types of configurable elements: input/output (I/O) blocks, logic blocks, and an interconnect matrix (see Figure P-1). From these, a designer can define individual I/O blocks that interface to external circuitry. You can think of these as configurable pins of ports. The designer can also use logic blocks, connected together through the interconnect matrix, to implement logic functions. These functions can be as simple as a counter or as complex as a microcontroller core. In a way the interconnect matrix is like the wires on a breadboard that connect everything together, but completely programmable.

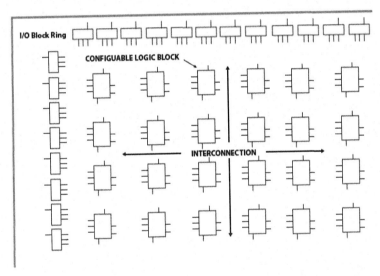

Figure P-1 *Xilinx LCA architecture*

Before there were FPGAs, you needed to use dozens of discrete ICs on a circuit board, or sometimes even hundreds of ICs on multiple circuit boards, to accomplish the hardware functionality you can achieve today with one FPGA device. For example, today you can create the entire Pac-Man game on a single FPGA device, including the game software. Now that's fun!

Xilinx and Altera are the two major players in the FPGA product space. Each of them provides a complete solution including a design tool suite. Altera came on the scene in 1992 when it introduced its first FPGA device family, the FLEX 10K line. There are pros and cons for each manufacturer, and many design wins come down to preference and price. For the purposes of this book, we will be focusing mainly on Xilinx, but the designs and experiments that follow will easily map to comparable Altera FPGAs, if you so desire.

The configuration of an FPGA device is accomplished through programming the memory cells, which determine the logic functions and interconnections. In the early days, the program (or what has become known as the *bit file*) was loaded at power-up from EEPROM, EPROM, or ROM on the circuit board, or loaded from a PC through a serial connection on the board from the FPGA programming tool. Since the underlying technology is volatile static RAM (SRAM), the bit file must be reloaded with every power cycle of the device. Today, SD flash memory replaces the EPROM and USB or JTAG replaces the serial connection, but the programming function remains much the same as it was in the beginning.

About the Book

This book is made up of eight interesting FPGA projects that will help you develop some of the skills you will need to really begin exploring this exciting world of turning software into hardware through FPGA technology. The projects will show you how to select an FPGA development board and set it up, and then give you the applications you will need to start making. The first chapter provides an overview of the boards and workflow we will be using in the book. The next two chapters walk you through a couple of simple projects, providing you with a hands-on look at the basics of the FPGA design flow. You can find the example code for these projects at my GitHub repository (*http://github.com/tritechpw/Make-FPGA*). The rest of the book concentrates on fun FPGA SoC projects. You can also check out my website (*http://tritechpw.com*) for more information on learning with FPGAs. Always remember that learning is a lifelong adventure. I hope you enjoy the journey.

Conventions Used in This Book

The following typographical conventions are used in this book:

Italic
> Indicates new terms, URLs, email addresses, filenames, and file extensions.

`Constant width`
> Used for program listings, as well as within paragraphs to refer to program elements such as variable or function names, databases, data types, environment variables, statements, and keywords.

`Constant width bold`
> Shows commands or other text that should be typed literally by the user.

`Constant width italic`
> Shows text that should be replaced with user-supplied values or by values determined by context.

This element signifies a tip, suggestion, or general note.

 This element indicates a warning or caution.

Using Code Examples

Supplemental material (code examples, exercises, etc.) is available for download at *https://github.com/tritechpw/Make-FPGA*.

This book is here to help you get your job done. In general, if example code is offered with this book, you may use it in your programs and documentation. You do not need to contact us for permission unless you're reproducing a significant portion of the code. For example, writing a program that uses several chunks of code from this book does not require permission. Selling or distributing a CD-ROM of examples from Make: books does require permission. Answering a question by citing this book and quoting example code does not require permission. Incorporating a significant amount of example code from this book into your product's documentation does require permission.

We appreciate, but do not require, attribution. An attribution usually includes the title, author, publisher, and ISBN. For example: "*Make: FPGAs* by David Romano (Maker Media). Copyright 2016 David Romano, 978-1-457-18785-8."

If you feel your use of code examples falls outside fair use or the permission given above, feel free to contact us at *bookpermissions@makermedia.com*.

Safari® Books Online

 **Safari Books Online** *is an on-demand digital library that delivers expert content in both book and video form from the world's leading authors in technology and business.*

Technology professionals, software developers, web designers, and business and creative professionals use Safari Books Online as their primary resource for research, problem solving, learning, and certification training.

Safari Books Online offers a range of plans and pricing for enterprise, government, education, and individuals.

Members have access to thousands of books, training videos, and prepublication manuscripts in one fully searchable database from publishers like O'Reilly Media, Prentice Hall Professional, Addison-Wesley Professional, Microsoft Press, Sams, Que, Peachpit Press, Focal Press, Cisco Press, John Wiley & Sons, Syngress, Morgan Kaufmann, IBM Redbooks, Packt, Adobe Press, FT Press, Apress, Manning, New Riders, McGraw-Hill, Jones & Bartlett, Course Technology, and hundreds more. For more information about Safari Books Online, please visit us online.

How to Contact Us

Please address comments and questions concerning this book to the publisher:

Make:
1160 Battery Street East, Suite 125
San Francisco, CA 94111
877-306-6253 (in the United States or Canada)
707-639-1355 (international or local)

We have a web page for this book, where we list errata, examples, and any additional information. You can access this page at *http://bit.ly/make-fpgas*.

Make: unites, inspires, informs, and entertains a growing community of resourceful people who undertake amazing projects in their backyards, basements, and garages. Make: celebrates your right to tweak, hack, and bend any technology to your will. The Make: audience continues to be a growing culture and community that believes in bettering ourselves, our environment, our educational system—our entire world. This is much more than an audience, it's a worldwide movement that Make: is leading—we call it the Maker Movement.

For more information about Make:, visit us online:

Make: magazine: *http://makezine.com/magazine*
Maker Faire: *http://makerfaire.com*
Makezine.com: *http://makezine.com*
Maker Shed: *http://makershed.com*

To comment or ask technical questions about this book, send email to *bookquestions@oreilly.com*.

Acknowledgments

Many experiences led me to ultimately writing this book. Some were years in the making as I made my way through a very exciting and rewarding engineering career. Along the way, many people helped me learn and gave me great opportunities to explore and invent. I can't list them all here, but I would like to thank them. I also want to thank my publisher, Brian Jepson, for offering me this chance, and my editor, Roger Stewart, who was so understanding throughout the writing process. I'd also like to thank Jack Gassett for all his help and support. Most of all I want to thank my wife, Elaine, for always believing in me and my Lord Jesus Christ for blessing me each step of the way.

Oh, and I can't forget Gracie the high-tech cat. She worked tirelessly by my side every day, overseeing the entire project.

Figure P-2 *Gracie, the high-tech cat, hard at work!*

Overview 1

When it comes to off-the-shelf FPGA development boards, the sky is the limit. There are many options to choose from. Prices can range from under $50 to thousands of dollars. For this book, I chose to keep the price window between $50 and $200. After price, one of the primary things to consider when choosing an FPGA board is what you will be using it for. For most of us it will be general experimentation, but some of you may already have a specific project in mind.

When looking at FPGA boards we obviously want to know the vendor, family, and size of the FPGA device. A good rule of thumb is the larger the device, the more it will cost. The next thing to look at is what features the platform supports. The most important of these is external interface connectors. Obviously, if you can't connect anything to the board easily, you are very limited in what you can do with it. It's good to note the size (number of pins) and frequency rating (this is less important for us) of the interfaces provided. Also note if the interfaces allow for connection to standard-form-factor add-on modules like the popular Ardunio shields, Digilent's Pmods, and the Papilio Wings. Other interface options to consider are the type and number of onboard standard communication interfaces that are provided, such as USB, Gigabit Ethernet, HDMI/DVI, PCI/PCI Express, external nonserial memory (DDR/flash, etc.), SD card, I2C/SPI, VGA, UART, etc.

FYI

Xilinx is standardized on the FPGA Mezzanine Connect (FMC), which is an industry standard. There are a lot of companies that make FMC-based plug-in cards.

I also like to have some number of onboard LED, switches, buttons, or even a 7-segment display, which are all very handy to have on the baseboard. These features can help you with the initial bring-up of the module, as we shall see in the next chapters, and also with the learning curve of the platform. They can also provide a convenient and cost-effective way to build some very basic experiments without the hassle of ordering additional boards or breadboarding your own circuits.

The other big thing to consider is the development tools that you will need to design and program your FPGA with. For our purposes, the tool license should be free of charge and provide a robust design entry method that supports both schematic capture and HDL (VHDL and Verilog) input. The tool should provide some basic simulation capability and a solid synthesis engine with good output reporting. Connection to the FPGA device should be easy with a standard interface like USB, and uploading to the FPGA should be done easily through a PC. OS support should include Windows and Linux with Mac as an option.

One other thing to consider on the FPGA platform is if there is some type of microcontroller integrated onboard or onchip. The Papilio DUO and Xilinx Zynq are examples of these. Having a local microcontroller available opens up another great dimension for creativity and exploration. In Appendix A, you will find a list of many low-cost FPGA boards that you can choose from.

In the rest of this chapter, I will review a few boards that I think have some unique features. The features that I'll be highlighting may provide you with some interesting opportunities for innovation.

Papilio

For the FPGA hobbyist and DIYers, Papilio by Gadget Factory is second to none. Everything about these products, from the array of affordable hardware modules to the innovative design environment and abundant learning material, was created with you in mind. As you can tell, I'm over the top with these guys. I felt like a kid on Christmas morning when I opened the box the day my shipment of Papilio products arrived. There were so many options to play with, I really didn't know where to begin. Now that's impressive!

So what is Papilio?

Papilio is a series of FPGA development boards and add-on hardware application modules called "Wings" that plug into the main board—sort of like Arduino shields.

As the Gadget Factory website points out, Papilio is Latin for butterfly:

> Papilio conveys the ability of an FPGA to undergo "digital metamorphosis." FPGA technology allows the Papilio to become any type of digital circuit including microcontrollers and custom chips such as the Commodore 64 audio (SID) chip.

You clearly get a sense of the creativity level that we are dealing with from that. This is going to be fun!

Papilio FPGA boards come in a range of affordable options, from the $37.99 Papilio One to the $87.99 Papilio DUO. You read that right: you can get into experimenting with FPGAs for as little as $37.99.

Did You Know?

With the $37.99 Papilio One 250K, you can start experimenting with FPGAs using the Xilinx XC3S250E, a 5.5K logic cell device. You also get onboard Papilio Wings connectors, a two-channel USB connection, and 48 general-purpose input/output (GPIO) pins.

I will start at the other end of spectrum, with the $87.99 Papilio DUO. This module is unique because it combines an Arduino-compatible microcontroller with a Xilinx FPGA on the same board. Not only does the board provide the ability to connect to the optional Papilio Wings modules, it also provides full compatibility with the Arduino shield ecosystem through the standard onboard Arduino connectors. There's even support for a "Pmod" (Peripheral Module interface, an open standard defined by Digilent Inc. in the Digilent Pmod™ Interface Specification for peripherals used with FPGAs or microcontrollers) connector on the board. Talk about add-on options!

The Papilio DUO uses the Xilinx Spartan-6 XC6SLX9 FPGA as its core. Table 1-1 describes the features of this FPGA device.

Table 1-1 *Spartan-6 XC6SLX9 features*

XC6SLX9 feature	Description
9K logic cells	Logic cell ratings are intended to show the logic density of one Xilinx device as compared to another device. Like a logic block, the typical cell includes a couple of flip-flops, multiplier, logic gates, and a small amount of RAM for a configurable lookup table (LUT). The logic cell is normalized to a 4-input lookup table (LUT).
576 Kb of block RAM	A block RAM (BRAM) is a dedicated two-port memory block containing several kilobits of RAM. The FPGA contains several (or many) of these blocks. In the Spartan-6 family, the block size is 18 Kb and the 6SLX9 has 32 of these, so the total size is 576 Kb.
2 CMTRs	There are two clock management tiles (CMTs) in the 6SLX9. Each CMT contains two digital clock managers (DCMs) and one phase-locked loop (PLL). The DCM core is a very versatile and complex piece of Xilinx intellectual property (IP). It can be used to implement a delay locked loop, digital frequency synthesizer, a digital phase shifter, or a digital spread spectrum.

XC6SLX9 feature	Description
TMDS I/O	Transition minimized differential signaling (TMDS) I/O support means that DVI and HDMI interfaces can be implemented directly with the FPGA I/O pins without any extra chips.
16 DSP slices	There are 16 digital signal processor (DSP) slices in the 6SLX9.

You can check out the full data sheet for the Xilinx Spartan 6-LX9 on the Xilinx website (*http://bit.ly/1SfM4iy*).

Opal Kelly

I chose to feature Opal Kelly (*https://www.opalkelly.com*) in this book because of their unique approach to bridging the gap between FPGA design/development and interconnection with a PC or other computing device using USB or PCI Express. For the hobbyist, this opens some very interesting opportunities for innovation. Opal Kelly offers a full product line of FPGA modules ranging in price from the very affordable XEM6001-6002 at $174.95 to the $1399.95 XEM5010-50M256. We will be using the XEM6002 for our experiments. This module uses the same Xilinx Spartan-6 XC6SLX9 FPGA as the Papilio DUO for its core. This FPGA contains 9K Logic cells, 576Kb of Block RAM, 16 DSP slices, and two CMTs. For details on this device, see the previous section.

The Opal Kelly XEM6002 module is also equipped with four Digilent Pmod-compatible connectors, as seen in Figure 1-1, which allow for interfacing to a wide variety of low-bandwidth peripheral modules available from several semiconductor manufacturers. Think of the possibilities!

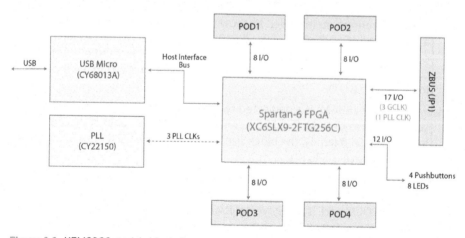

Figure 1-1 *XEM6002 module block diagram*

Check out the Digilent Peripheral Modules page (*https://www.digilentinc.com/Products/Catalog.cfm?NavPath=2,401&Cat=9*) for a look at some of the fun and interesting plug-in modules you'll have access to.

Did You Know?

Pmods are small I/O interface boards that offer a great way to extend the capabilities of an FPGA or embedded controller board. Pmods communicate with system boards using standard 6- or 12-pin connectors. In addition to various sensors and connectors, there are Pmods for I/O, data acquisition and conversion, external memory, and much more.

The key to the Opal Kelly innovation is the FrontPanel SDK. The software development kit, or SDK, provides a small FPGA library block that integrates with your FPGA design to make USB (or PCI Express) host communication simple and easy. It also includes a software API, simplifying the programming development of the communication interface, and a robust driver to communicate with your device over USB or PCI Express (see Figure 1-2). The USB driver and FrontPanel API work together to provide an easy-to-use software interface to your hardware that is consistent across the Windows (32-/64-bit), Linux (32-/64-bit), and Mac OS X development environments.

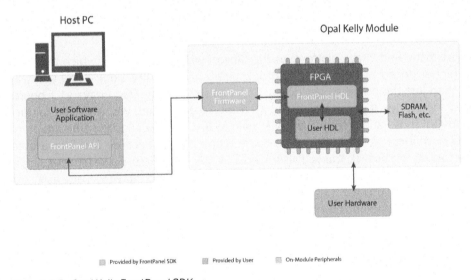

Figure 1-2 *Opal Kelly FrontPanel SDK*

Opal Kelly also provides prebuilt wrappers to the FrontPanel API for C, C#, C++, Python, and Java, and the DLL can be used from any software that allows external calling, such as MATLAB or LabVIEW.

The standalone FrontPanel application lets you quickly and easily define your own graphical user interface (GUI) that communicates with your hardware. It supports a variety of user interface elements, including LEDs, hexadecimal displays, sliders, push buttons, checkboxes, toggle buttons, and numerical entry. You can think of the FrontPanel as a virtual breadboard environment.

Wow, A Virtual Breadboard Sandbox!

With the Opal Kelly FrontPanel virtual interface you can quickly and easily create a GUI of an instrument right on your PC that interfaces with your FPGA design! How cool is that?

Red Pitaya

The Red Pitaya platform (*http://www.redpitaya.com*), according to its developers, is an "open-source measurement and control tool replacing many expensive laboratory instruments." It is based on the Xilinx Zynq-7010 (*http://bit.ly/1K7IqEl*) programmable SoC device, which combines a dual-core ARM® Cortex™-A9 MPCore processor with a programmable logic array that contains 28K logic cells, 240 Kb of block RAM, and 80 DSP slices. The processing subsystem operates autonomously from the programmable logic, "booting on reset like any other processing device." It acts as the "system master" and controls the configuration of the programmable logic, enabling full or partial reconfiguration during operation.

The retail cost of the Red Pitaya platform at the time of writing is $238.80, which makes it one of the higher-priced platforms of the group being used for this book—and you will need more than just the board to get going. Figure 1-3 shows the Red Pitaya v1.1 board and required accessories. If you are planning on using scope probes, you will need to purchase a couple of those and a couple of SMA male to BNC female adapters (the adapters are required to connect standard BNC-type oscilloscope probes because Red Pitaya uses SMA connectors on the board). You will also need a 5V, 2A, USB micro, power supply, and a 4–8 GB micro SD card. All of this is conveniently offered on the Red Pitaya website as a Red Pitaya website as a Diagnostic Kit (*http://store.redpitaya.com/red-pitaya-board-32.html*) for a grand total of $310.80, not including any taxes or shipping.

Figure 1-3 *Red Pitaya board and accessories*

Did You Know?

The ARM A9 core is the application processor used in many tablets and cell phones. It is a high-performance, low-power microprocessor core.

What is really impressive about this platform are the three major technology subsystems that you have access to. The first is the RF/analog frontend, which gives you the ability to instantiate a true oscilloscope, spectrum analyzer, or signal generator. The second is the ARM CPU core, and the third is the FPGA itself. You can purchase premade instrument files (Oscilloscope + Signal Generator or Spectrum Analyzer) from the Red Pitaya online store (*http://store.redpitaya.com*). Check out the Detail specification (*http://bit.ly/1KKcf8Z*) page on the Red Pitaya wiki for a full description of the frontend hardware.

That being said, let's crack open the box and see what's inside (Figure 1-4). Like on a first date, first impressions are lasting ones. I must admit I was impressed by the small size of the platform, measuring only about 4 × 2.5 inches (the website calls it a "credit card footprint").

Figure 1-4 *The impressive small size of the Red Pitaya platform*

Numato Lab

Numato Lab (*http://www.numato.com*) provides a complete product line of low-priced FPGA development boards and accessories. But just because the prices are low doesn't mean that they are short on features—on the contrary, the bang for your buck you get with these products is amazing. I was able to take a look at three of the Numato FPGA platforms, pictured in Figure 1-5:

- The $29.95 Elbert V2 - Spartan 3A FPGA Development Board

- The $49.95 Mimas V2 Spartan 6 FPGA Development Board with DDR-SDRAM
- The $199.95 Waxwing Spartan 6 FPGA Development Board

Figure 1-5 *Numato Lab Elbert V2 (top), Mimas V2 (middle), and Waxwing (bottom)*

That's right—for just $29.95 you can jump into working with FPGA technology! I may sound like a used car salesman, but this is not a barebones FPGA board. Take a look at the Elbert V2 feature list:

- FPGA: Spartan XC3S50A in TQG144 package
- 16 Mb SPI flash memory (M25P16)
- USB 2.0 interface for onboard flash programming
- FPGA configuration via JTAG and USB
- 8 LEDs, six push buttons, and an 8-way DIP switch for user-defined applications
- VGA output
- Stereo audio out
- Micro SD card adapter
- Three 7-segment displays
- 39 I/Os for user defined-purposes
- Onboard voltage regulators for single power rail operation

This is a remarkable value, as you get almost everything you need: LEDs, push buttons, DIP switches, and even three 7-segment displays. The only downside is that the Spartan XC3S50A is a 1.5K logic cells device, compared to the Spartan-6 LX9 FPGA, which is a 9K logic cell device.

If FPGA size is an issue, for another $20.00 you can jump up to the Mimas V2, which has a Spartan-6 XC6SLX9 and also includes 512 Mb of DDR memory along with all the other features of the Elbert.

For $199.95, the Waxwing development board should really be called a development laboratory. This board has everything you can imagine on it:

- Spartan-6 FPGA (XC6SLX45 in CSG324 package on Waxwing Mini Module)
- 166 MHz 512 Mb LPDDR
- 100M Ethernet (LAN8710A)
- National Semiconductor LM4550 AC '97 audio codec
- 128 Mb SPI flash memory (W25Q128FV)
- 100 MHz CMOS oscillator
- DVI-D connector for video (compatible with HDMI)
- 8-bit VGA output connector
- 2 channel audio out through 3.5 mm audio jack
- 16 × 2 character LCD display
- Micro-SD adapter
- 3 common anode 7-segment LED displays
- 7 onboard push button switches
- 44.5 × 35.1 mm mini breadboard for easy prototyping
- High-speed USB 2.0 interface for onboard flash programming
- FT2232H channel A dedicated for SPI flash programming; channel B can be used for custom applications
- FPGA configuration via JTAG and USB
- Onboard voltage regulators for single power rail operation

They even provide you with a mini-solderless breadboard, right on the board. This is really an experimenter's dream—you don't need to buy any additional add-on modules to do some really cool experiments.

Design Flow

For all of these boards, the design process is basically the same, so I think this is a good place to provide a short overview of a typical FPGA design flow. Most SoC projects follow a similar path when creating a new device, as illustrated in Figure 1-6. The process usually starts with a concept, transitions to the actual design of the device, followed by testing, synthesis, building, and finally running or executing the device in the system.

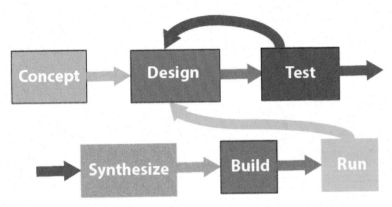

Figure 1-6 *SoC design flow*

Each phase contributes to the overall success or failure of the project.

Concept Phase

You can see that we start with the concept phase for any given SoC project. The concept may be just an idea or something more formal, like a design description or even a written functional specification. In the concept phase, you're pretty much free to dream up anything, but you will also need to do some preliminary feasibility analysis to see if what you are imagining will actually map to the technology you are targeting. In our case, it's FPGA. You need to consider things like number of I/O pins and speed, amount of on-chip RAM required, and a rough estimate of size (gate count) and clock speed, to name a few. In designs where you haven't already selected an FPGA, you have the luxury of doing the design first and then synthesizing it. From the synthesis reports you can see how many FPGA resources the design requires, and you can then select the optimum FPGA device for your particular design. In most cases, though, this will not be an option because you will already have purchased the FPGA board. Instead, you will need to adapt your design to your FPGA device. Most of the time when you are dealing with IP block libraries, the description provides a gate count number. You can get a quick ballpark estimate by just doing some research on the IP blocks you think you will be using in your design and adding up the gate counts. This is where a quick sketch of a block diagram for your design will be extremely handy.

Design Phase

For me, the design phase is the most fun. There are multiple ways to approach SoC design, but I'm going to describe two basic methods. The first is the *top down* method of design. This is where you start with a very high level view of the design and work your way down into the details. For me, this is the most natural method, and it's the one we will be using most in our upcoming experiments. This method is also known as the *hierarchical design* method. The block diagram that you sketched in the concept phase should be a good representation of the top level of your design hierarchy. Design hierarchies can be represented in HDL form, graphical form, or both. For example, you could have a top level that looks something like the block diagram in Figure 1-7.

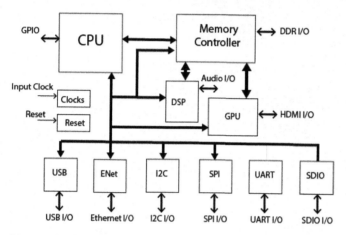

Figure 1-7 *Top SoC level*

Drilling down one level into the I2C (Inter-Integrated Circuit) block, the diagram could look something like Figure 1-8.

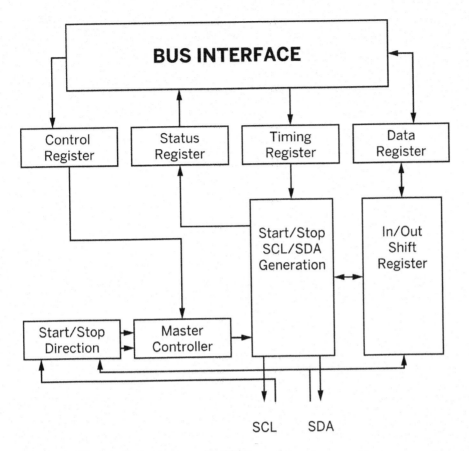

Figure 1-8 *I2C controller top level: graphical form*

Or it can be represented in Verilog as shown in Figure 1-9.

```
module i2c_master_top(
  wb_clk_i, wb_rst_i, arst_i, wb_adr_i, wb_dat_i, wb_dat_o,
  wb_we_i, wb_stb_i, wb_cyc_i, wb_ack_o, wb_inta_o,
  scl_pad_i, scl_pad_o, scl_padoen_o, sda_pad_i, sda_pad_o, sda_padoen_o );

  // parameters
  parameter ARST_LVL = 1'b0; // asynchronous reset level

  //
  // inputs & outputs
  //

  // wishbone signals
  input         wb_clk_i;     // master clock input
  input         wb_rst_i;     // synchronous active high reset
  input         arst_i;       // asynchronous reset
  input   [2:0] wb_adr_i;     // lower address bits
  input   [7:0] wb_dat_i;     // databus input
  output  [7:0] wb_dat_o;     // databus output
  input         wb_we_i;      // write enable input
  input         wb_stb_i;     // stobe/core select signal
  input         wb_cyc_i;     // valid bus cycle input
  output        wb_ack_o;     // bus cycle acknowledge output
  output        wb_inta_o;    // interrupt request signal output

  reg [7:0] wb_dat_o;
  reg wb_ack_o;
  reg wb_inta_o;

  // I2C signals
  // i2c clock line
  input  scl_pad_i;      // SCL-line input
  output scl_pad_o;      // SCL-line output (always 1'b0)
  output scl_padoen_o;   // SCL-line output enable (active low)
```

Figure 1-9 *I2C controller top level: partial Verilog code example*

Each subblock in the hierarchy is another logical function of the next level of design detail. In this way, a very complex design is broken down into very manageable pieces. With HDL, it is easy to stitch together the hierarchy and create a useful file structure to manage and control the design. This is an example of the benefits and power of using HDL, and it's how you can create a design that is made up of over a billion transistors. With HDL and hierarchical design you can efficiently break the design down and have teams of engineers working on different blocks at the same time. There can be hundreds or even thousands of engineers working on a single SoC design at once. There are hardware engineers doing this very thing, right now, all over the world.

The other design method is called the *bottom up* method. Here you start with one section of your design and work out all the details of that aspect of the design, and then move to the next section. This method is also known as the *flat design* method. Schematic drawings are typically used for a flat representation of a design. There can be many sheets of schematics tied together, but they are all at the same very detailed level. This method works well for very small designs, but when you get into larger SoCs, the hierarchical method is really the only the way to go. Most printed wiring boards (PWBs) are flat designs and use computer-aided design (CAD) schematic entry tools to capture the design, but flat design is not typically used in SoC development.

Test Phase

The test phase, also known as the validation phase, is basically where you apply some type of stimulus to the design or a design element and observe the results. This sounds simple enough, but SoC validation can be just as or even more complex than the design itself. In industry, the validation team can be two or three times the size of the design team for a complex SoC project. This is particularly the case for full custom or standard cell chips. Remember the cost is extremely high for that first chip with these technologies, so you need to make every effort to be sure that it is right and more than just pass the DOA test.

FYI

A DOA (dead on arrival) test is a very popular test for SoC validation engineers to write (typically it is the first test they write for a new chip or design block). This test usually checks to see if the chip or design has any life at all after power-up. For example, does the chip come out of reset to a known state? If the DOA test fails, there is something catastrophically wrong with the design.

The test bench is the method used for most SoC validation. The test bench is nothing more than a simulation wrapper that is put around the unit under test (UUT), along with virtual test generators and test instruments to stimulate and monitor the UUT (see Figure 1-10). For SoC designs, the test bench can be written in HDL or C++. Behavioral models and bus functional models are used as generators and monitors in the test bench. A behavioral model is HDL code that mimics the operation of a device, like a CPU, but is not gate-level accurate. In other words, it is not synthesizable. When coding bus functional models you have much more freedom to optimize out many details that don't affect the focus of your test—these models only create the sequence of states on a bus necessary for the test (for example, read/write transactions on the CPU bus). (In the last decade, a language called System Verilog has become very popular in industry validation engineering circles, but it's beyond the scope of this book and we won't be covering it here.)

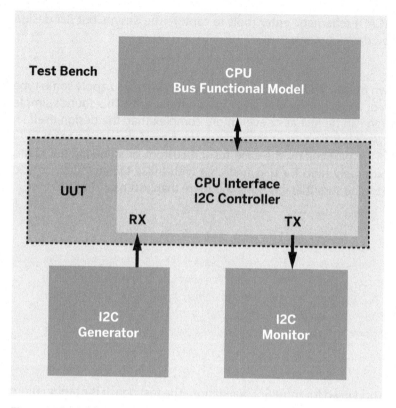

Figure 1-10 *I2C controller test bench*

A test can also be as simple as a quick simulation of your design, conducted by manually forcing the clock and control signals from within the simulation tool itself, then visually inspecting the output waveforms for the results. With this method no extra test bench code is required. I like to use this method to check my design while I'm in the design process itself—which brings up another interesting point. There are basically two design-validation philosophies that one can follow. The first is the complete design method. Here the designer carefully maps out and plans every detail of the design block, then codes it in HDL, writes the test bench code, and then tests it. There is very little iteration in this method with the exception of fixing bugs in the design code. The second method is the iterative method of design. Here the designer builds up the design in stages of complexity iterating between design and test. There could be many iterations between the design phase and test phase in this method until the design converges on the expected output results (see Figure 1-11).

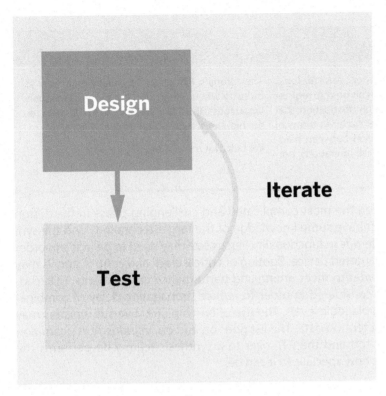

Figure 1-11 *Design-test phase iteration*

There are obviously pros and cons for both of these methods and much of it comes down to style and personal preference. I tend to use the iterative method more because I like to use the power of the simulation tools to help me visualize my design assumptions in real time. Keep in mind that there is always some number of iterations between design and test phases; nothing works perfectly the first time through!

Synthesize Phase

Logic synthesis is the process by which the register-transfer level (RTL) of the SoC is turned into a design implementation in terms of logic gates, typically by a CAD tool called a *synthesis tool*.

RTL

Not to be confused with register transistor logic, the term used in transistor circuit design, register transfer level (RTL) is a design abstraction that models a synchronous digital circuit in terms of the flow of digital signals (data) between hardware registers and the logical operations performed on those signals.

The acronym RTL is used by many engineers in industry when referring to hardware description languages (HDLs) like Verilog and VHDL that create high-level representations of a circuit.

We talk a lot more about RTL in Chapter 2.

The synthesize phase is often the most complicated and challenging phase in the design flow, and the designer must have some knowledge of the target technology and the synthesis tool being used. Trade-offs in functionality versus size may need to be made in order for the design to fit in the targeted device. Routing of critical clock and control signals may need to be considered in order to meet timing and performance requirements. HDL coding styles may need to be considered in order to reduce propagation delay of combinational signals through multiple logic levels. The use of on-chip SRAM versus registers may to be considered for state retention data. The list goes on and on. In industry, it's common for the hardware engineer to hand the RTL over to a synthesis team who performs the actual synthesis run—that's how specialized it can be.

In our case, the FPGA tool from Xilinx that we will be using has a pretty good synthesis tool included, but the old adage "garbage in, garbage out" reigns supreme. As we go through each experiment, I will be covering different aspects of the synthesis process, so we will be coming back to this topic in upcoming chapters.

Build Phase

The build phase can be as simple as a push of a button, as is the case for us with FPGA and CPLD (complex programmable logic device) technologies, or can entail a multitude of manufacturing steps, costing many thousands or even millions of dollars (as is the case for some of the other technologies we discussed earlier in this chapter). We won't be getting into integrated circuit manufacturing in this book. The build phase for us just involves hooking up our USB cable to our FPGA module and uploading the bit file to the FPGA device. Piece of cake!

Run Phase

The run phase is where the design is put into operation, physically executing in the target technology. Here, the proof is in the pudding, although with FPGA this is not as critical as it is with the nonreprogrammable technologies. For us, if the design does not work we will need to go back through the flow and begin debugging. Often, in simple FPGA designs, the validation phase can be pushed into the run phase. Testing is done only at the power-on, chip level. For example, if your FPGA design just blinks an LED, there is no real reason to build a test bench for that. For that matter, there is no real reason to do a simulation of

---

the design. Just by running the design you will know whether it works or not. That is one of the big advantages of designing with FPGAs—they are reprogrammable!

As illustrated in Figure 1-12, our toolbox of integrated development environment (IDE) tools will focus on for the most part the design, test, synthesize, and build phases of an SoC project.

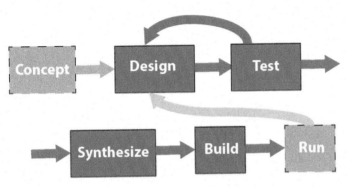

Figure 1-12 *Focus of IDE design flow*

Takeaways

The main takeaway for this chapter is that there are many FPGA development boards to choose from, in a very wide price range (for a more comprehensive list of low-cost options, see Appendix A). As I said at the beginning of the chapter, it really comes down to what features you are looking for and what your budget is. The good news is that there is something for everyone. Here are a couple of other points to keep in mind:

- Most of the boards are easy to set up on Windows, but for the most part Linux support is definitely lagging and Mac support is not in the mix. You may just have to struggle through the setup with many of these boards, because I have found that no matter how smoothly it may go on one machine, on another it can be a complete nightmare. It's just part of the joy of DIY!

- When we talk about FPGA design, a lot of people immediately get all wrapped around the axle thinking you have to be a VHDL or Verilog guru in order to jump in. That is far from the truth—in fact, one of the main purposes of this book is to prove that you don't have to be a VHDL or Verilog expert to start building cool FPGA projects!

Count on It! 2

Basic FPGA Design Flow Using a Frequency Divider Circuit

In this chapter, I'll walk you through a basic FPGA design flow using a simple frequency divider circuit. I recommend that everyone read through this chapter just to review the basics. If you are new to FPGAs and FPGA design methods, this will be a very practical guide for you. If you have some experience designing with FPGAs, then you may find this to be a good refresher.

I'll be going through, in some very detailed steps, installing the Xilinx ISE tool, creating a project in the Xilinx ISE tool, creating a schematic and Verilog source file, synthesis, basic simulation, creating a user constraints file (UCF), bit file generation, and programming your target FPGA device. I will not be going through these steps in the following chapters, because I'll assume you have mastered them here before moving on.

When I begin working with any new FPGA module, I like to start with a very simple design to get me familiar with the lay of the land, so to speak, of the module and tool flow. I use a very simple counter circuit to help me do this. We'll be following the SoC development flow outlined in the previous chapter and shown again in Figure 2-1 as our guide to building this simple test circuit.

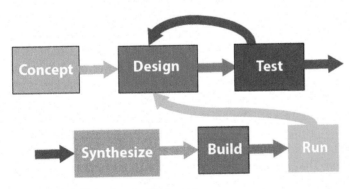

Figure 2-1 *SoC development flow guide*

Keep in mind that this little circuit gives you the opportunity to quickly go through the whole design process, touching on all the major concepts and functions of your module. You'll be able to flush out a lot of the ins and outs of the design by going through this experiment. I highly recommend that you start here.

Blink LEDs Concept

The requirements for our design are very simple: we would like to blink two indicator lights (LEDs) on and off at different rates based on the system clock of our FPGA module. We would also like to have a mechanism to hold the circuit in reset. A nice-to-have requirement is for the lights to be different colors, say red and green. We can sketch a simple block diagram for the system as shown in Figure 2-2.

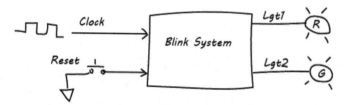

Figure 2-2 *Blink system sketch*

How It Works

The theory of operation for this system is simple. Our input clock will be operating at some nominal frequency—let's say 100 MHz. We want to use the clock signal to somehow flash our two LEDs (lights). We know that frequency is equal to the inverse of the period, so 100 MHz is equal to a 10 ns period. If we turn the LED on and off every 10 ns, though, it will be way too fast for our eyes to register the transition. We won't see the light blinking at all, even though it will actually be blinking very rapidly. We need to divide the primary clock signal down to a much slower frequency in order for our eyes to register the on-to-off state transition of the LED. In essence, we want to generate a slower clock from a faster clock. We can observe that when counting in binary, the least significant bit toggles twice as fast as the next significant bit. The same is true of the next significant bit, and so forth. Given this fact, we can use a simple binary counter as a clock frequency divider circuit (see Figure 2-3). The question now is, how many bits does our counter need to be? We know that at flicker rates beyond 100 times per second, humans stop noticing the blackness of, say, a TV screen refresh. House lights turn on and off at 60 Hz, the frequency of electric power, and our eyes cannot detect the flicker. The LEDs here are similar. We can think of each as one TV screen pixel. So, we have to slow the 100 Mhz clock down below the 100 Hz range. We need to divide our 100 Mhz clock frequency by at least a million to get our eyes to notice the change from off to on. That means we need at least a 20-bit binary counter (2 raised to the 20th power).

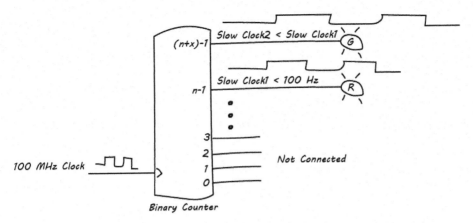

Figure 2-3 *Clock frequency divider circuit*

Xilinx ISE WebPACK Installation

Now is a good point to talk about the development tools we will be using. Just as any good builder needs a toolbox full of the right tools for the job, the FPGA designer needs a good toolbox with the right tools to complete an FPGA project. The primary tool in our FPGA design toolbox will be the ISE WebPACK from Xilinx. First and foremost, ISE is free, and you can install the tool on as many computers as you wish. For students and hobbyists, I think free tools are a must-have! You still need to generate a license file with the install, which enables only the free features of the tool. The feature set that Xilinx provides with the free license will be more than sufficient to meet our needs.

As of October 2013, ISE has moved into the "sustaining" phase of its product lifecycle. But this does not mean that it will no longer be available for use by the Make community—on the contrary, it is my understanding from my sources at Xilinx that ISE will be available for the foreseeable future; there just will not be any new features added to it. For all new Xilinx device families, starting with the Artix® family, you will need to use the Vivado® Design Suite WebPACK (*http://bit.ly/1KKhG84*) from Xilinx. While this WebPACK edition is also free, it has a much more limited feature set than ISE. We will not be using Vivado in this book.

Getting started

You will need to create a Xilinx account to proceed with the download. First, you will need to go to the Xilinx website and download a copy of the ISE WebPACK that matches the OS of the computer you will be running it on. Choose Support→Downloads Licensing in the main menu bar, and then click the ISE tab (*http://bit.ly/1JWwCVt*); the default download page is for Vivado, and you don't want to download that by accident! The ZIP file is very large, and it may take a while for it to complete the download.

The Xilinx website also provides you with the option of requesting a free DVD version (*http://bit.ly/1KKhJAI*) of the software, which can be shipped to you. You can request only one copy of the DVD.

Installation

The installation of the ISE WebPACK is relatively straightforward. You can find the official install guide (*http://bit.ly/1KKhLIU*) on the Xilinx website. They also provide a good tutorial guide (*http://bit.ly/1KKhNR3*) on the tool.

Here's a quick summary of the installation process:

1. Unzip the download file to a directory on your computer.

2. Run the *xsetup.exe* file. For Linux installs, you will need to run as a superuser (*sudo*).

3. Choose the defaults and accept the licensing agreements.

4. Under "Product to select," choose ISE WebPACK.

5. You can select the defaults for everything else.

6. Wait for the "Install Completed" message.

7. Once the install is complete it will pop up a window where you can select Acquire a License. Select "Get Free Vivado/ISE WebPack License" and click Next.

8. You should automatically be connected to the Xilinx page, and a license will be generated for you. Sometimes that may not work, though, and you will have to generate, download, and install the license manually. To install the license manually:

   a. Go to the Xilinx website (*http://xilinx.com/getlicense*).

   b. Log in to the account you created to download the installer.

   c. Click Manage Licenses and find your WebPACK license. Click the Download icon in the bottom-left corner to download your license.

   d. Go back to the License Configuration Manager and click the Load License button. A file dialog should pop up, and you will need to select the license file you just downloaded.

9. Run ISE WebPACK. Linux users will need to run the start script (*settings32.sh*, located in the */opt/Xilinx/14.7/ISE_DE* directory) first, then type **ise** on the command line.

Did You Know?

You can install the ISE WebPACK on as many computers as you like—just copy the same license file you generated to the other computers. You do not need to generate a new license file for each computer.

Once you have the Xilinx ISE WebPACK installed on your computer and have gone through your FPGA modules and start-up guide, you are ready to take that first big step on your journey of FPGA design.

Design

With our system concept in hand, we can now start our design. The first thing we'll need to decide is what method we will use to capture our FPGA design details. For our first design we will use graphical schematic entry. I'll also show the HDL version of the same design later in the chapter. The graphical entry will enable us to see some of the features of the design easily, especially if you're new to this.

We'll also need to build a small breadboard circuit that will act as our peripheral device, which will contain our LED lights that will connect to our FPGA module. I like to build this external circuit because it is completely independent of any LEDs on your module. In this way you really get to dig into the I/O mapping of your board. You can also add a push button for the reset function or just use a jumper wire.

Peripheral Breadboard

If you're like me, you'll have all of the components required to build this little circuit lying around your bench somewhere; if not, take a look at the Bill of Materials (BoM) for a parts list.

The BoM for the circuit shown in Figure 2-4 will include a couple of LEDs, current limiting resistors, a pull-up resistor, a switch (optional), some male-to-male jumper wires, and a small solderless half-size breadboard.

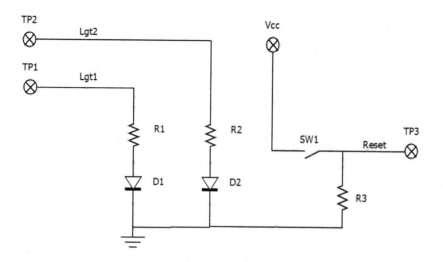

Figure 2-4 *Blink LED peripheral schematic*

BoM

R1, R2, R3 = 330—10K ohm (any value of resistor in this range will work)

D1 = Red LED (can be any color)

D2 = Green LED (can be red or any color)

TP1, TP2, TP3 = Connection points in your breadboard

SW1 = 1 × push-button switch (optional)

1 × small 1/2 size breadboard

4 × male-to-male jumper wires

Once you have all the parts, it should only take you a few minutes to build this circuit on your breadboard. It should look something like Figure 2-5.

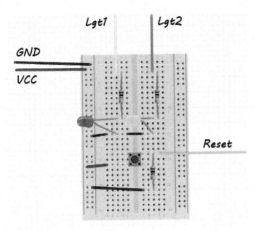

Figure 2-5 *LED breadboard*

FPGA Circuit Schematic Design Entry

Now we get into the good stuff! Let's open our first, new, FPGA project in the Xilinx ISE WebPACK tool, as seen in Figure 2-6.

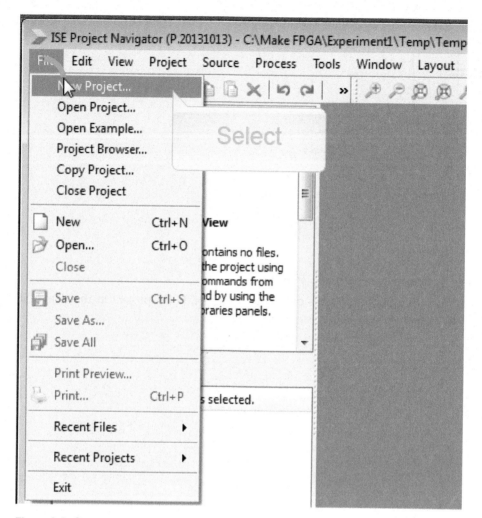

Figure 2-6 *Open new project*

Name the new project *BlinkLEDs1*. Don't forget to select Schematic as the top-level source type, as shown in Figure 2-7.

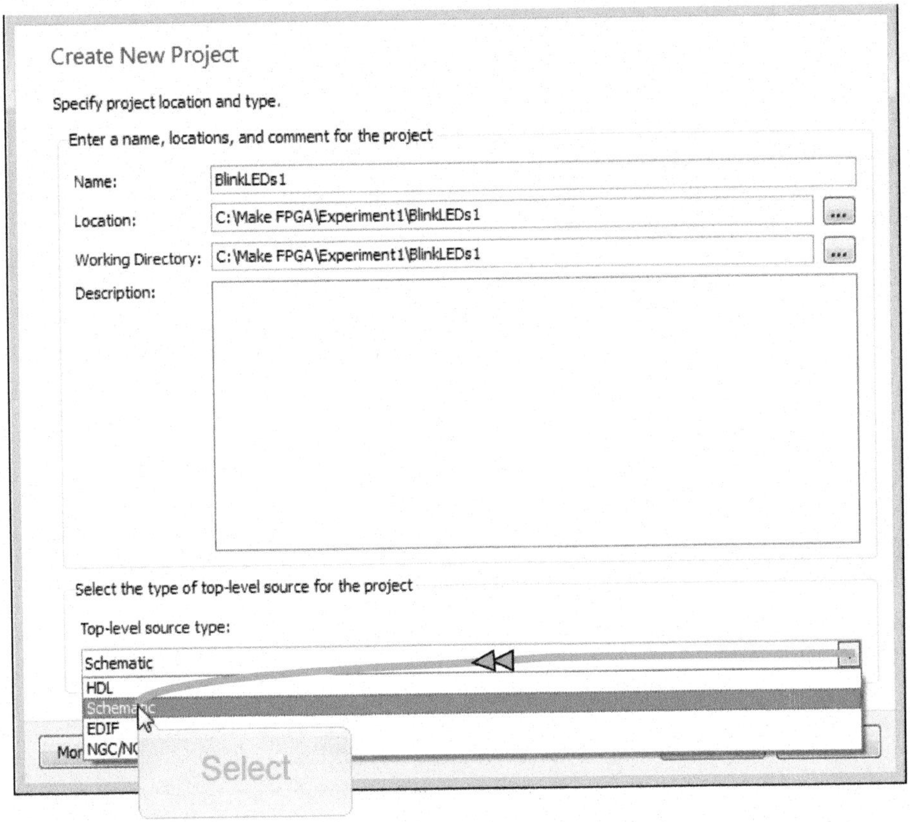

Figure 2-7 *Name the project and select Schematic as the top level*

You should also specify your device properties at this point, as seen in Figure 2-8. The family, device package, and speed are all dependent on the FPGA module you are using.

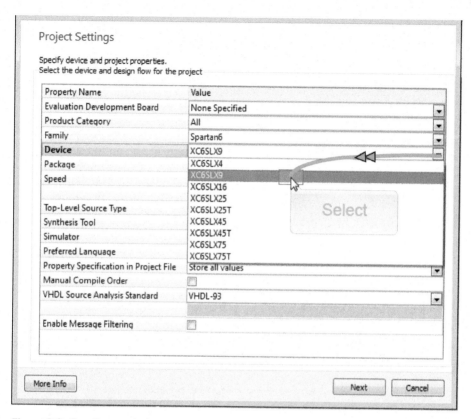

Figure 2-8 *Specify your device properties*

Heads Up!

Be sure to specify the correct FPGA device information for your module! You will need to have these properties set correctly for later, when you build. Do it now before you forget! It will save you a whole lot of errors down the road.

You can click through the defaults for the rest of the settings at this point.

Next, we will open a new source file, as shown in Figure 2-9.

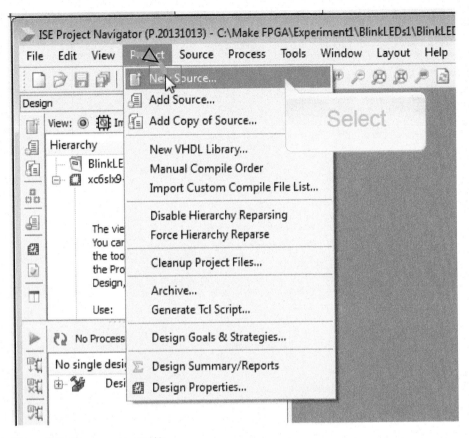

Figure 2-9 *Open new source file*

Select Schematic as the source type, name the file *BlinkSystem*, and hit Next, as seen in Figure 2-10.

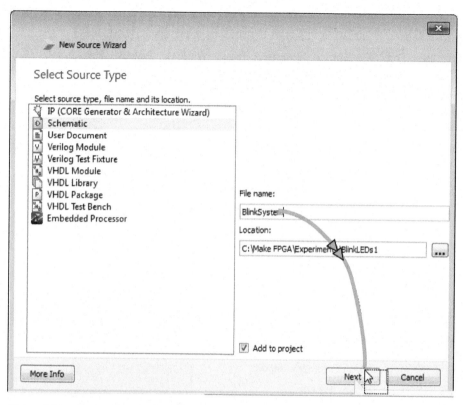

Figure 2-10 *Name your schematic file and hit Next*

We're almost there! Now go to the Design panel and click your *BlinkSystem.sch* file (see Figure 2-11). This will be your schematic drawing.

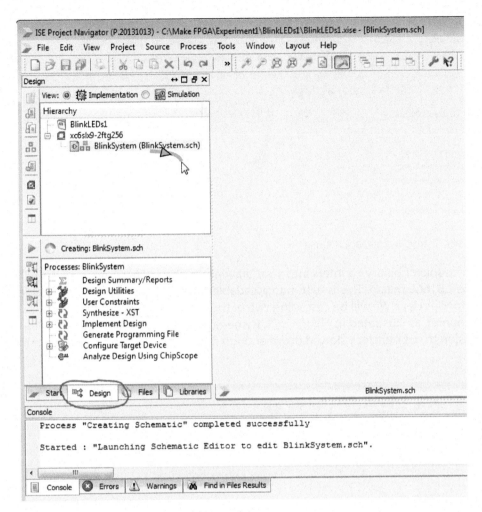

Figure 2-11 *Open the schematic drawing*

We are now ready to start designing our circuit using the schematic symbol library. You may want to zoom in a bit on your drawing at this point (see Figure 2-12).

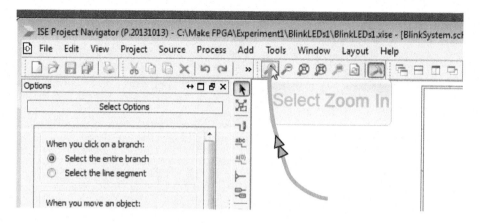

Figure 2-12 *Zoom in to your schematic drawing*

Now drop a couple of binary counters into your drawing, as shown in Figure 2-13. We will be using the CB16CE macro. This is a 16-bit cascadable binary counter with clock enable and asynchronous clear. We will be cascading two of these macros together to create a 32-bit binary counter, as illustrated in Figure 2-14. Remember we need at least 20 bits of frequency division to get our clock slowed down enough for our eyes to see the difference in on/off time.

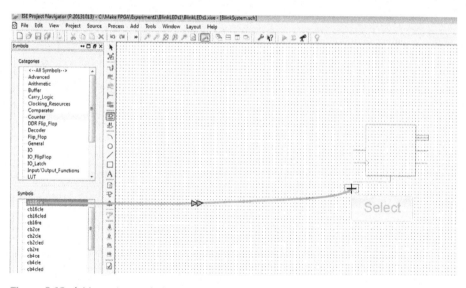

Figure 2-13 *Add counter symbol*

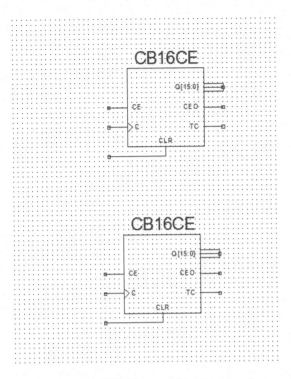

Figure 2-14 *Two CB16CE cascadable binary counters*

 FYI

Xilinx provides a set of guides that describe all the macros in the schematic library for different device families, such as the Spartan-6 Libraries Guide for Schematic Designs (http://bit.ly/1K7lsMV).

Now we are ready to wire up our counters. We'll start by clicking on the Add Wire tool in the toolbar on the left (see Figure 2-15).

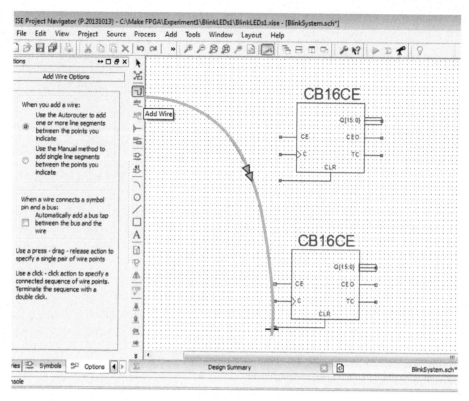

Figure 2-15 *Add a wire*

Connect the two CLR inputs together with wires, then connect the two C (or clock) inputs together. Now add a single wire to the bottom CE (count enable) input, but don't wire this to the top one. Next, connect the two Q(15:0) output buses together. Finally, connect the bottom CEO (count enable out) to the CE of the top counter. This is how you cascade the counters together. When the lower counter reaches its terminal count, it will generate a single output strobe, for one clock cycle, on its CEO pin. This will be used to enable the high count, so for every 2^{16} lower counts the high count will count once. Your schematic should look something like Figure 2-16.

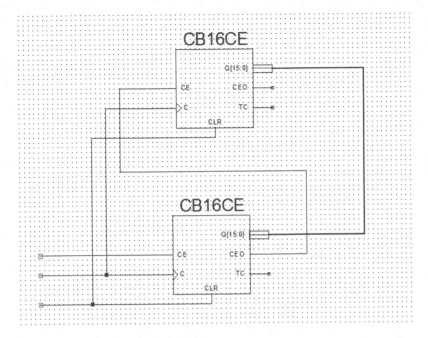

Figure 2-16 *Wire up your counters*

Now let's add some labels to our wires and buses. It's very important to use a naming convention for labeling—this is a good design habit to get yourself into. For small designs like this, it is not critical, but in larger designs or joint designs it becomes essential. I like to use lowercase text for internal signals and buses. Standard convention is to use either the "_n" or "_low" designation at the end of the signal name for signals that are active low. Figures 2-17 and 2-18 illustrate the procedure for adding a label.

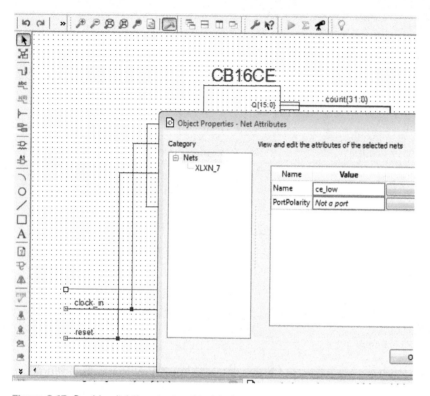

Figure 2-17 *Double-click the wire to add a label*

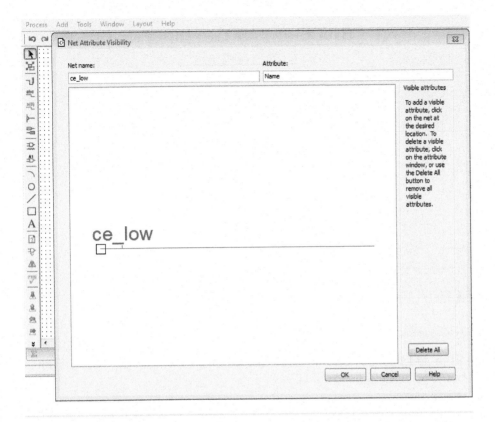

Figure 2-18 *Click Apply and then Add to add text, then click OK*

When you're done, you should have labels that look like those in Figure 2-19. Notice that the output bus is labeled "count(31:0)." This is because this is a 32-bit (wire) bus; 31:0 means 31 down to 0.

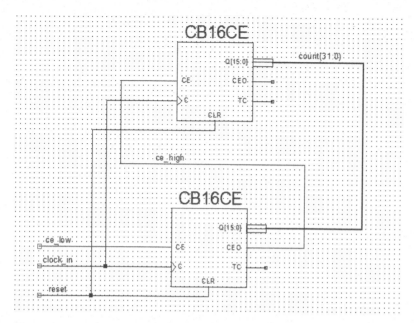

Figure 2-19 *Signal names added*

Next, we need to add bus taps on our 32-bit output bus, so we can get two individual wires of this bus to act as outputs to our LEDs.

Clicking the Add Bus Tap icon in the lefthand toolbar brings up the Add Bus Tap Options pane in the Options panel, as seen in Figure 2-20.

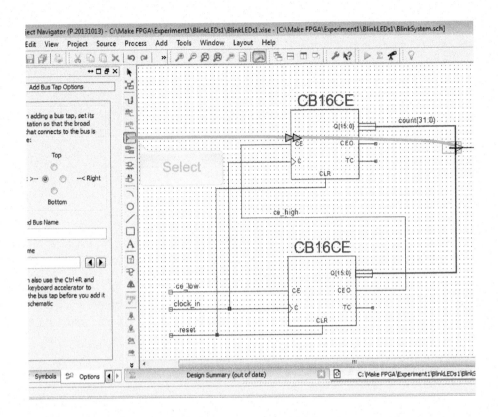

Figure 2-20 *Add bus tap*

Notice that the Add Bus Tap tool's Options panel provides an orientation selection. You want the bus tap's V-shaped end to intersect perpendicular to the bus that you are trying to tap. In this example, the tap requires a left orientation in order to connect to the bus properly.

The next step is required for the build process and is mostly just a quirk of ISE. We need to add an output buffer between the bus tap and the output pin, which we will be adding next. If you fail to do this, you will not be able to name your output pin correctly. Select the *obuf* from the *IO* symbol library category, add it to your schematic, and connect the bus tap and obuf together with a wire (see Figure 2-21). Do this for both counter signals.

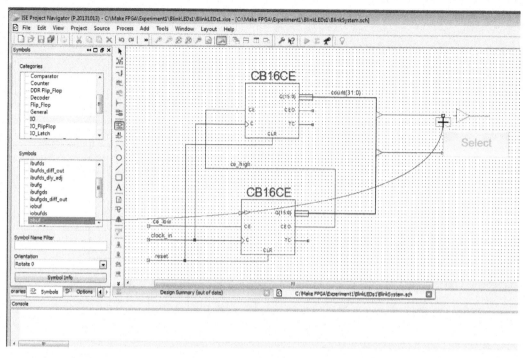

Figure 2-21 *Add an obuf*

Add labels to your wires and you should have something that looks like Figure 2-22. I chose to assign bits 20 and 21 of the 32-bit counter to the two bus taps. This will be a good place to start our experiment because we know that we need a reduction of at least 1 million (20 bits) for our 100 Mhz input clock.

Heads Up!

Don't forget to occasionally save your work as you create your schematic!

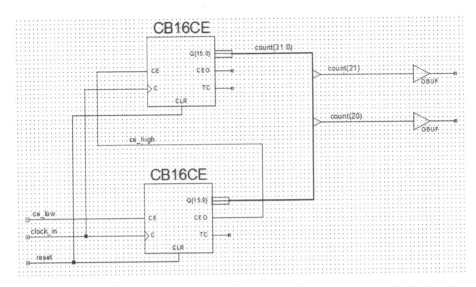

Figure 2-22 *Schematic with bus taps added*

The last thing we need to add to our schematic is I/O pin symbols. These symbols will tell the ISE Place and Route tool (discussed in "Build") what physical FPGA pins we want our signals to be active on. We also need to add a pull-up function to our *ce_low* signal. We need the lower half of our 32-bit counter to count continually on every input clock cycle. To achieve this, we will simply assign a constant value of 1 (high) to this input signal. This is done in the schematic editor by placing a pull-up symbol on the wire, as seen in Figure 2-23. I also named the I/O pins with capital letters, keeping to my convention. This allows for easy identification of physical I/O pins from internal signals when simulating.

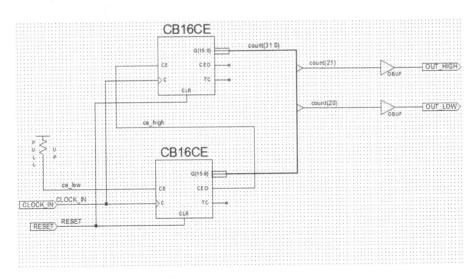

Figure 2-23 *Complete schematic*

You are now ready to implement (compile) your module. To do this, simply click the green triangle on the top toolbar. You should get the following errors when you do this:

- ERROR:DesignEntry:20 - Pin "Q(15:0)" is connected to a bus of a different width.
- ERROR:DesignEntry:20 - Pin "Q(15:0)" is connected to a bus of a different width.

I intentionally constructed our schematic to generate these errors to illustrate the point that the Xilinx schematic editor is not very robust in many regards. This is another one of those quirks that is a real pain to deal with. Bus merging, apparently, is something that the ISE schematic tool does not do very well. To work around this problem, I found that we need to break the 32-bit bus into two 16-bit segments, giving each segment a unique name (see Figure 2-24). Keep in mind that this is not a design requirement; it is a tool work-around. As you grow in your FPGA design capabilities, you may encounter similar errors that turn out to be other tool issues. When you encounter an error, click the word "ERROR" in the error message string in the console window. This will bring up some information about the error in the Xilinx database. You can also search the Web for information that relates to the problem by cutting and pasting the error message into your search engine or searching on a related topic, such as "Xilinx schematic bus merging."

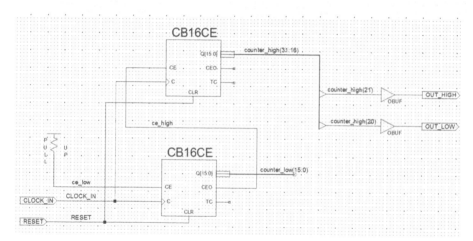

Figure 2-24 *Schematic with errors corrected*

Once you correct the schematic and re-run the compile, which will take a few seconds, you should get a "Process 'Generate Post-Place & Route Static Timing' completed successfully" message in your console window. You should get in the habit of reviewing any warning messages that are generated. In this case, there should be a few of them, but a quick review shows that the warnings pertain to the unconnected *counter_low* bus and are of no concern. When debugging a problem, it is also a good idea to go back and look at the warning reports for clues.

FPGA Circuit HDL Design Entry

Now we'll capture the same design using an HDL source file instead of a schematic. This will give you a good opportunity to experience firsthand the benefits of using an HDL editor to capture your design details. For this example, I will be using Verilog to code the design. At this point, some of you may be thinking HDL stands for "hard/difficult language" instead of "hardware description language," but here you'll see just how HDL really simplifies SoC design.

To get started we will create a new project in ISE and name it *HDL_BlinkLEDs1*. Be sure to select HDL as the top-level source type when you create your new project, as shown in Figure 2-25.

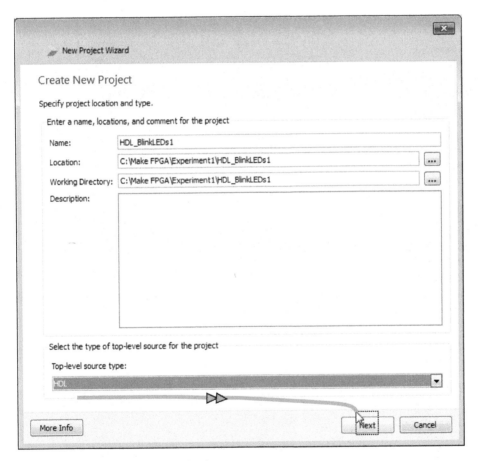

Figure 2-25 *Select HDL as the top-level source*

After you click Next, select Verilog as your preferred language in the Project Settings dialog box (see Figure 2-26).

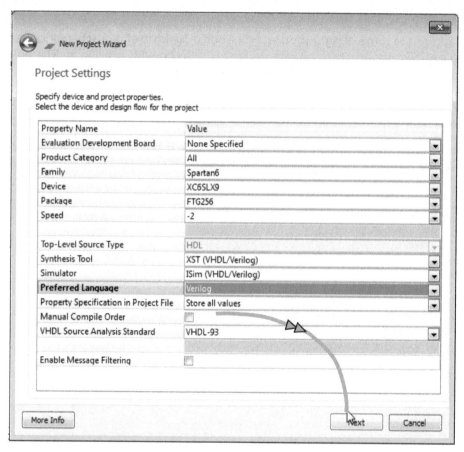

Figure 2-26 *Preferred Language to Verilog*

Hit Next, and open a new source file from the Project menu in the top toolbar. Select Veri-log Module from the Select Source Type dialog box and name your file *BlinkLED_Sys2*, as shown in Figure 2-27.

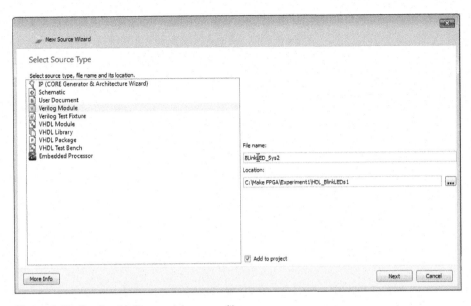

Figure 2-27 *Create a Verilog module source file*

After clicking Next again, you will need to define your module's ports. Verilog modules are like functions in other programming languages. They are pieces of code that can be used and reused within a single program. Verilog modules are defined using the *module* keyword, and end with the *endmodule* keyword. The module is declared with a module name (in this case, *BlinkLEDSys2*), followed by a list of parameters, which is the list of input and output signals that connect to the module. When we synthesize our Verilog code, the parameters will signify the physical wire connections that connect our module to the rest of the SoC.

Think of the module as a top-level block in a block diagram. Remember our system block diagram from earlier in the chapter (Figure 2-2)?

This is the module we are describing now in Verilog. Notice that the block has four ports: two input ports labeled Clock and Reset and two output ports labeled Lgt1 and Lgt2. The great thing about the ISE tool is that it gives you a headstart on your source file construction with a handy port declaration wizard. For this example, we can just rename the labels to CLOCK_IN, RESET, OUT_HIGH, and OUT_LOW using all capital letters, keeping to our naming convention (see Figure 2-28).

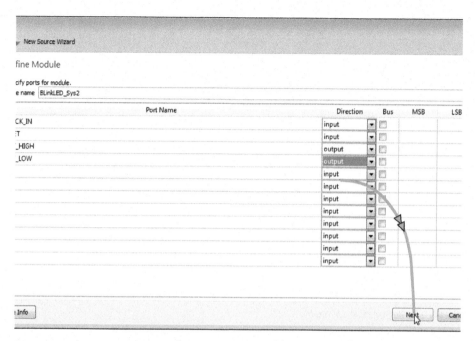

Figure 2-28 *Port declaration wizard*

Click Next and just like magic, you've created your first HDL source file (Figure 2-29)—but we're not done yet!

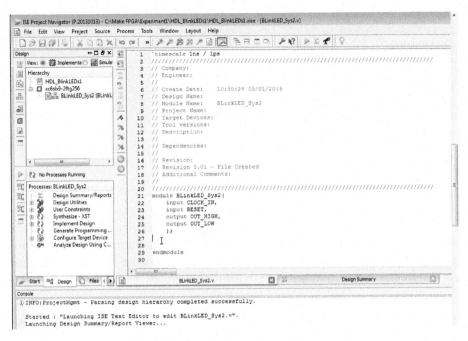

Figure 2-29 *Verilog module source file*

Notice that the ISE HDL editor color-codes the text. This is very helpful when constructing and reading code. First, you'll see there is a lot of green text. This text is comments, and it has no effect on the compilation of the code. You can add comment text anywhere in the source file by simply preceding the text with two forward slashes (//); this tells the compiler that what follows on this line is a comment. You can add a multiline comment by starting each line with //, or you can use a /* (forward slash and asterisk) at the beginning of the comment and a */ (asterisk then forward slash) to close your comment block. You can use this method to easily comment out a whole block of code when you are debugging, for example.

Full details on the color coding for Verilog and VHDL are given in Table 2-1.

Table 2-1 *ISE text color coding*

Color	Verilog	VHDL
Green	Comment	Comment
Blue	Reserved word	Reserved word
Pink	Directive	STD LOGIC reserved word
Orange	Function	N/A
Red	UNISIM reserved word	UNISIM reserved word
	SIMPRIM reserved word	SIMPRIM reserved word

Color	Verilog	VHDL
Gray	Attribute string	String
Black	Default text Identifier Number Operator	Default text Identifier Number Operator

Most of the time you will be coding with blue reserved words and black identifiers.

In Verilog, circuit components are designed inside the module. Modules can contain both structural and behavioral statements. Structural statements represent circuit components like logic gates, counters, and microprocessors. Behavioral-level statements are programming statements that have no direct mapping to circuit components like loops, if-then statements, and stimulus vectors that are used to exercise a circuit.

Our next step is to code our 32-bit counter circuit component. Basically, a counter is just a code loop that adds one to itself each time through the loop. This is illustrated in pseudocode in Example 2-1.

Example 2-1 *Pseudocode of counter*

```
Start
     count = count + 1.
   goto Start
```

Since Verilog is a hardware description language, we need to code up some of the basic hardware building blocks of our circuit—in this case, registers and wires. If you don't have a hardware background, this may seem a little confusing at first, but let's go back to our counter diagram. We said that we needed at least a 20-bit counter to meet our requirements and in our previous schematic example, we used a 32-bit counter. To keep everything the same, we will use a 32-bit counter in our HDL example, too. The 32-bit counter needs to be realized in hardware by 32 storage elements called *registers* or *flip-flops*. Flip-flops are digital logic circuits that can be in one of two states. They maintain their state indefinitely until an input pulse called a *trigger* is received. Typically this is the rising edge of an input clock. When a trigger is received, the flip-flop outputs change state according to the defined rules and remain in those states until another trigger is received. Flip-flops can be used to store one bit of data. The data may represent the value of a counter, the state of a sequencer, an ASCII character in a computer's memory, or any other piece of information.

Did You Know?

The most common flip-flop is the **D-Flip-Flop** (DFF) or **D Register**. As illustrated in **Figure 2-30**, this is represented by the schematic symbol that has two inputs and one output: C (clock), D (data in), and Q (data out). You can see the DFF in your ISE schematic symbol library by choosing "fd" in the "Flip_Flop" category. The DFF transfers the value on D to Q on the rising edge of C.

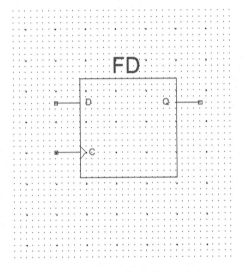

Figure 2-30 *ISE DFF schematic symbol*

So, we need to define our counter variables to be made up of 32 registers (DFF.) We do this by using the reg reserved word, followed by the size [31:0] and the name blinkcount:

```
reg[31:0]blinkcount;
```

Notice the semicolon at the end of the text string. Verilog requires a semicolon at the end of every code string. If you get errors when you first compile your code, check for proper semicolon placement. You will most likely miss one or two while coding your first pass, which will generate a lot of bizarre errors.

Heads Up!

Make sure you add a semicolon at the end of every Verilog code string.

We also have to declare the wires that connect from our module port to our counter:

```
wire clk_in;
wire reset_in;
```

Now we have three elements defined in our circuit: ports, registers, and wires. We just need to hook them together, and we do that through the *assign* statement:

```
assign clk_in = CLOCK_IN;
assign reset_in = RESET;
assign OUT_HIGH = blinkcount[21];
assign OUT_LOW = blinkcount[20];
```

Notice that I did not define a wire from our counter but tied the blinkcount bit 21 and bit 20 directly to the ports OUT_HIGH and OUT_LOW. We can do this because Verilog understands that the output of a register is an internal wire or signal. When using the assign statement, you must be aware of the direction of the signal flow as it relates to the equal sign. The flow in the declaration is from right to left, so in our assign clk_in = CLOCK_IN statement we are assigning the input port CLOCK_IN to the wire clk_in. You can say that the wire clk_in "gets" port CLOCK_IN. Notice the difference in the last two declarations—the output of our counter blinkcount[21] is going to port OUT_HIGH or in other words OUT_HIGH gets blinkcount[21]. In this case, the port is on the left side of the equals sign because it is at the end of the flow.

Now we just need to define our counter loop and reset functionality:

```
always @(posedge clk_in)

if (reset_in) begin
  blinkcount <= 32'b0;
end

else

begin
  blinkcount <= blinkcount + 1;
end
```

We start by declaring that we want all state changes to always happen on the rising edge of the input clock. In other words, this is a synchronous design. We next use the behavioral if statement to perform our reset function. What we are saying here is that if our signal reset_in is true (high), then all the bits of the counter, 31 down to 0, are equal to zero. Then we use our else statement to begin counting on every rising edge of our clk_in signal, which is connected to our CLOCK_IN port, when reset_in is not true (low).

FYI

The CLR pin of the CB16CE used in the earlier schematic design (see
Figure 2-14*) is an* ***async reset*** *(i.e., reset) or the clear function happens asyn-*
chronously to the clock. In this Verilog code I use a ***sync reset****, meaning the*
reset is not active until the rising edge of the clock. It's a subtle difference
for this design and really doesn't make much of a difference, but it's some-
thing you should be aware of when you are designing.

The complete code for our counter circuit is shown in Figure 2-31.

```
19  //
20  /////////////////////////////////////////////
21  module BLinkLED_Sys2(
22      input CLOCK_IN,
23      input RESET,
24      output OUT_HIGH,
25      output OUT_LOW
26      );
27
28  //-----Internal Variables-----
29  reg[31:0]blinkcount;
30
31  //---- internal signals
32
33  wire clk_in;
34  wire reset_in;
35
36  //-------Code Starts Here------
37
38  assign clk_in = CLOCK_IN;
39  assign reset_in = RESET;
40  assign OUT_HIGH = blinkcount[21];
41  assign OUT_LOW = blinkcount[20];
42
43  always @(posedge clk_in)
44  if (reset_in) begin
45      blinkcount <= 32'b0;
46  end
47  else
48  begin
49      blinkcount <= blinkcount + 1;
50  end
51
52  endmodule
```

Figure 2-31 *Completed counter circuit Verilog code*

FYI

*You can find this code on GitHub (****https://github.com/tritechpw/Make-***
FPGA*).*

You are now ready to compile your code. Hit the green triangle and check for errors.

Next, I'll show you a cool tool that I like to use to see how the HDL synthesizer has interpreted my code. You are going to love this one!

Go to Tools→Schematic Viewer and select RTL (see Figure 2-32).

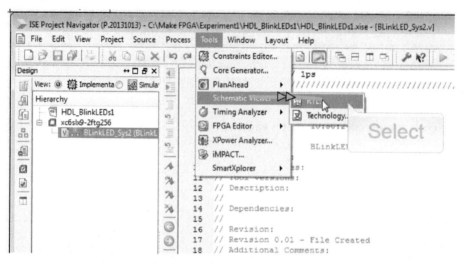

Figure 2-32 *Launch the RTL schematic viewer*

In the dialog box that appears (see Figure 2-33), select "Start with a schematic of the top-level block." I like to use this view because it builds the schematic using the register transfer level (RTL) abstraction model form we talked about in Chapter 1.

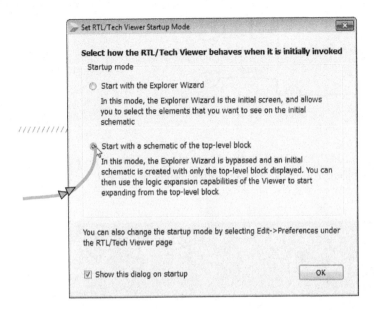

Figure 2-33 *Select top-level block*

Click OK and, just like magic, there is your top-level block! It should look like Figure 2-34.

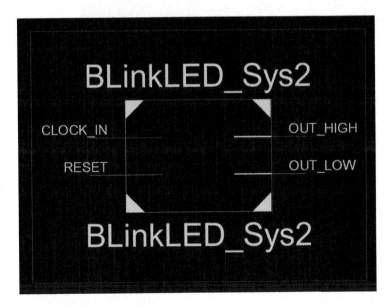

Figure 2-34 *Top-level block*

It's funny how it looks a lot like our system block diagram, reproduced in Figure 2-35. That was no accident.

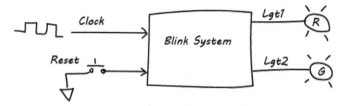

Figure 2-35 *Blink system block diagram*

Now, are you ready for more fun? Double-click the BlinkLED_Sys2 block and click "Zoom to Full View" in the upper toolbar. You now see the bottom level of the RTL view, as shown in Figure 2-36.

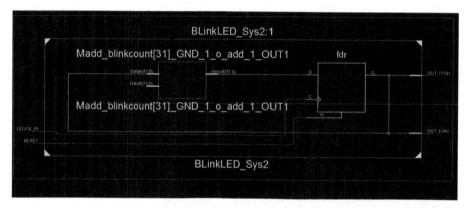

Figure 2-36 *Bottom level of RTL view*

You will notice two blocks here: the one on the right represents our 32 registers (reg[31:0]blinkcount), and the one on the left is what we call the combinational logic block. These two blocks represent the current state (right block) and next state (left block) logic sequences. Notice that the clock only goes to the right block. That is because it is registering (FF) the next state, which becomes the current state on every rising edge of the clock. Remember our always @(posedge clk_in) statement? This is how the synthesizer interpreted this and implemented it. Notice that the output of the fdr block (right) is fed back to the input of the left block. This is the next state generation logic. Remember when we said that blinkcount <= blinkcount + 1? Well, the synthesizer is doing just that in this block: it's taking the current count, which is the feedback from the registers (current state or current count), adding one to it in the left block and presenting that as the data going into the register block (right) to be the clock-in on the next rising edge of the clock (i.e., this becomes the new current count). This is a classic RTL model of how most synchronous HDL logic designs get coded and synthesized, represented graphically in Figure 2-37. If you can remember this simple RTL model, you will be able to understand a lot of HDL code.

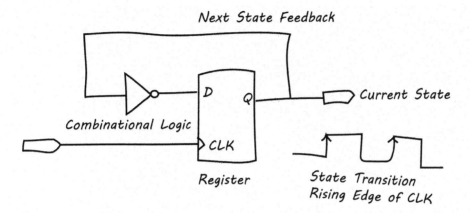

Figure 2-37 *Simple RTL design abstraction model*

We now can see that a synchronous circuit consists of two kinds of elements: registers and combinational logic. Registers (usually implemented as D flip-flops) synchronize the circuit's operation to the edges of the clock signal, and are the only elements in the circuit that have memory properties. Combinational logic performs all the logical functions in the circuit, and it typically consists of logic gates.

Now let's take a look at the bottom level of the Technology view (Figure 2-38). Select Tools→Schematic Viewer→Technology, double-click the BlinkLED_Sys2 block, and click Zoom to Full View in the upper toolbar.

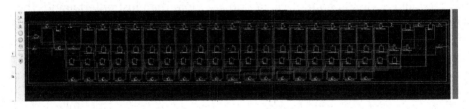

Figure 2-38 *Bottom level of Technology view*

Now we see the details of what the synthesizer actually built and is targeting for the FPGA. If you count the blocks in this view, across the schematic, you'll see that there are 21 blocks. Each one of these is a bit in our counter. You're probably thinking, "Wait a minute, I coded a 32-bit counter, not 21!" That's right, but since we tied only our high-order bit 21 to an output port, the synthesizer optimized out the unused bits. That's a good thing because it saves FPGA resources. We were able to easily verify this optimization with this view.

A closer look will show you some of the elements of our logic design. In the zoomed-in view in Figure 2-39, we can see the input buffer on our reset line, the clock buffer on our clock line, and some XOR gates, registers, and multiplexers (a multiplexer, or mux, is a digital logic element that selects one output out of several inputs).

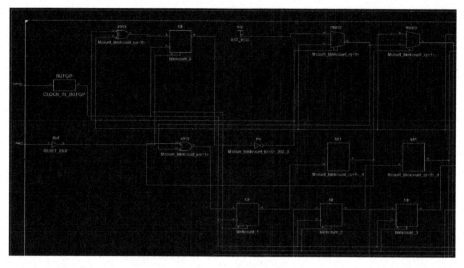

Figure 2-39 *Elements of logic design*

The option to generate HDL code from your schematic source file is also available in the ISE tool. To do this, go to the Design panel, select a *.sch* file in the Implementation window of your schematic design project, and then click "View HDL functional model." This will generate the HDL code for the selected schematic. But it won't be very useful because schematic-to-HDL conversion is typically very messy. You may use it to get a look at some basic HDL code, but it most likely will confuse you more than help. You've been warned.

The RTL and Technology views of your HDL design are more than just entertaining. They give you a good graphical way to check on how well your code is being interpreted by the synthesizer. As with everything, however, there is more than one way to skin a cat, and HDL coding is no exception. Some styles of coding result in a more optimum synthesis outcome (using less resources or less logic levels) while others tend to blow up on you and the synthesizer will create a lot of levels of logic. You can read the Xilinx whitepaper, "HDL Coding Practices to Accelerate Design Performance" (*http://bit.ly/1JW7FcH*) to fully explore this topic.

You may be wondering why all this is important. There are two answers. First, there are only a finite number of FPGA resources in your chip, so you'll want to use these in the most optimum synthesis solutions. The second reason is what we call "meeting timing." Remember in our bottom level of the RTL view (Figure 2-36), we had the register block and the combinational block. Getting all those combinational logic steps through all of those gates that you see in the Technology view can take a long time. This is called *propagation delay*, which is the total path delay for D to CLOCK IN. If you have a very high-speed clock, the cycle time between rising edges of the clock may not be enough time for the signal to propagate through all that combinational logic. For example, if we are using a 100 MHz clock, then the period from edge to edge is 10 ns. If the worst-case delay through the longest combinatorial path is 10 ns or greater, we have a problem. In this case your design may work functionally (in simulation) but will not work in the real world (in the FPGA). Timing issues are the most challenging issues to debug in an FPGA design, and we can spend a whole book talking about this topic. For our purposes, it's good to know that we will not be working with clock speeds that push the envelope of our FPGA technology. The ISE synthesis tools have come a long way over the years.

To help FPGA designers with analyzing their design, Xilinx has provided some timing tools in ISE that we can use to generate all kinds of timing reports. We'll take a quick look at some of these now, just so you'll know where they are if you need them later.

Start by clicking Tools→Timing Analyzer→Post Place & Route. This will activate the Timing tool in your toolbar and generate a Data Sheet report. This report provides the setup and hold timing information for your design. From the Timing tool, you can go to Reports and generate the Clock Topology, Reports by Clock Regions, and Nets by Delay reports. Of these, the one I find most useful is the Nets by Delay report. This gives us the 20 worst nets by delay. The Net report provides us with the net delay and not the total path delay, which can be affected by multiple net delays, logic delays, and FF setup and clock-to-out delays. You cannot determine directly if the design meets timing just by looking at this report. If you add a constraint file and put a constraint on the clock, you will be able to determine immediately if you are meeting timing. It's another step, but it can be crucial if the design is complex and pushes performance, which in our case is not a concern.

From the Net report (see Figure 2-40), you can quickly tell which net is going to make timing just by inverting the delay that will give you the maximum clock frequency at which a particular path can operate. If the frequency of operation is above your system clock, then you are in good shape. If the frequency of operation is below your system clock, then that particular net has a problem.

```
The 20 worst nets by delay are:
+-----------------+-----------+
| Max Delay       | Netname   |
+-----------------+-----------+
    3.662           RESET_IBUF
    3.239           blinkcount<21>
    3.189           blinkcount<20>
    1.494           CLOCK_IN_BUFGP
    0.907           CLOCK_IN_BUFGP/IBUFG
    0.460           blinkcount<19>
    0.460           blinkcount<3>
    0.460           blinkcount<15>
    0.460           blinkcount<11>
    0.460           blinkcount<7>
    0.405           blinkcount<16>
    0.405           blinkcount<12>
    0.405           blinkcount<0>
    0.405           blinkcount<4>
    0.405           blinkcount<8>
    0.237           blinkcount<13>
    0.237           blinkcount<5>
    0.237           blinkcount<9>
    0.237           blinkcount<17>
    0.237           blinkcount<1>
    -----------------------------------
```

Figure 2-40 *Net Delay report*

For example, in our sample report we see that the RESET_IBUF net has a max delay of 3.662 ns. If we invert this, we get 273 MHz as our maximum frequency of operation. If our system clock is 100 MHz, then we have plenty of headroom on this one net (more than double).

It's a good idea to keep this notion of "making timing" in the back of your mind during the final build step. This is when we will be able to run our "Post Fit" timing analysis and generate our final reports. Until we do this step, the timing information is just preliminary, so you need to be cautious when viewing the results. If you are not making timing before the fitting stage, it will most likely only get worse.

Simulation

Now we are ready to do a quick test of our circuit before we build it and load the bit file to our FPGA. I always like to try a simple simulation first, even for small designs, because it is a good sanity check to see if the design works at a rudimentary level. I think it is a good habit to get into, rather than just going right to building and loading, even if you just take the design out of reset in the simulation. You would be surprised how many bugs I have found in HDL code just by doing this. This will be a simple manual simulation test; we won't be coding an elaborate test bench for this exercise.

We'll start with our schematic design project first and then rerun the same simulation on our HDL design project. We expect the results to be the same. To start the simulation, select Simulation in the Design panel, select your schematic, and then click Simulate Behavioral Model. This will open the simulation window (see Figure 2-41).

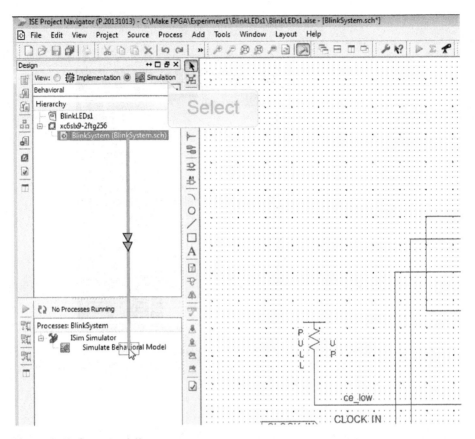

Figure 2-41 *Start simulation*

You will next need to force your clock and reset signals to known states in order to provide stimulus to your circuit, which will cause the simulator to respond or change state. You do this by right-clicking the label of the signal you want to set (force). Since the clock is a repeating periodic signal, we will use the "Force Clock" selection to set our CLOCK_IN parameters, as shown in Figure 2-42.

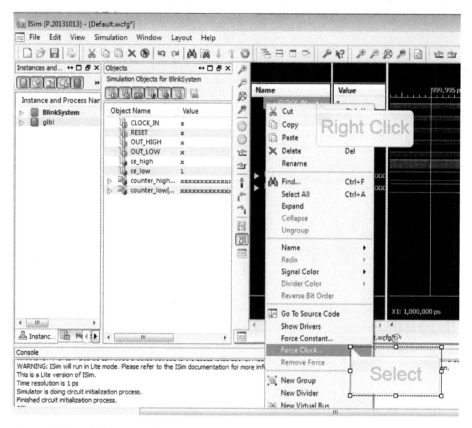

Figure 2-42 *Select force clock*

Set the leading edge value to 1, followed by the trailing edge value to 0, and the period to 100. Then hit OK.

We next will force our reset signal to 1, which will place our circuit in reset, clearing our counter. Right-click on the RESET signal and select Force Constant this time (we don't want a periodic signal here—we are just looking for a steady state signal). Set Force to Value to 1, as seen in Figure 2-43, and click OK.

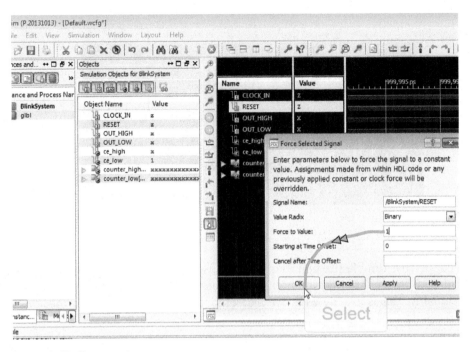

Figure 2-43 *Force reset*

Now select "Run for the time specified on the toolbar," which defaults to 1 microsecond (1.00 µs). Select Zoom to Full View, and you should see all the signals turn to green, the clock toggle, and the output of your counter set to all zeros (see Figure 2-44).

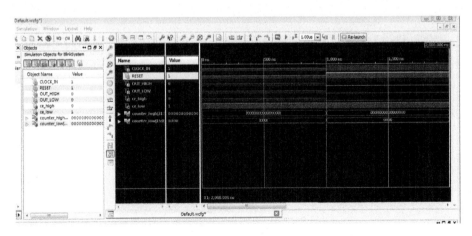

Figure 2-44 *Run for 1.00 µs*

Now force the RESET signal to 0 and change the time in the Run window to 1000 µs. You should see you OUT_HIGH and OUT_LOW outputs toggling now. You can zoom in and out to see your signals and counter outputs more closely (see Figure 2-45).

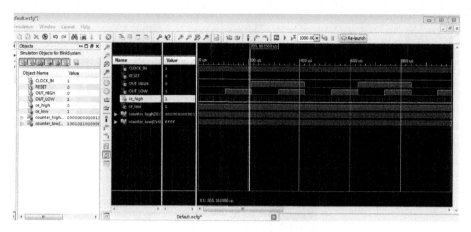

Figure 2-45 *Run for 1000 µs*

Notice how OUT_HIGH is double the period of OUT_LOW, which is because the higher we go in the bit order of our counter, the greater our frequency division is, and therefore the slower our signal is. This is exactly what we expected, and it appears that our design is functioning properly.

You can try doing this same simulation of your HDL design project, and you will see the exact same results (Figure 2-46). This confirms that the two design projects are functionally identical, just captured in two different ways.

FYI

Remember async versus sync resets? The sims may still look the same depending on when the reset is asserted in relation to the clock (see Figure 2-46).

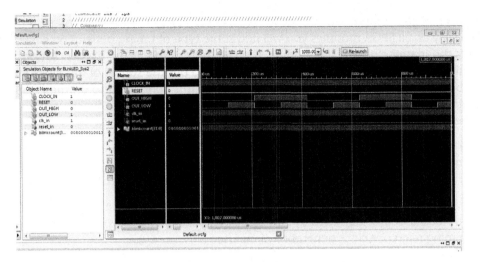

Figure 2-46 *HDL project simulation*

We are now ready to continue on to the next step in our design flow.

Build

In the build process, we will be assigning the physical I/O pins of our FPGA to our design by creating a constraint file. Then we will re-run the ISE implementation tool, which will perform the place and route processes with our pin assignment information.

Assigning Physical I/O

This next step can be the most difficult in the whole process, because it is here that you will need to link the virtual world of your ISE design to the real world of the actual FPGA chip and your particular FPGA module's circuit board layout. Sometimes it can take a little bit of detective work to pull all the pieces together. It really depends on how good your FPGA module vendor is at providing documentation; some do a better job than others. The good news is that once you go through it, you will have all the information you need for every build after that, so it's really only painful the first time through. The key to success is to know what you are looking for.

We'll start by going back to our original block diagram model (Figure 2-47).

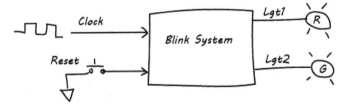

Figure 2-47 *Original block diagram model*

Notice that we need an input clock in order to operate our synchronous blink system design. Typically on all FPGA modules you will find some type of timing generation circuit that outputs the master clock for the module. You can think of this as the master platform clock or timing circuit. The first thing we need to do is find out the source of this clock on the circuit board and determine what physical FPGA pin it is connected to on the FPGA chip. A good place to start is the user manual of your board, if there is one available; if not, the next place to search is the schematic of the module, if one is available.

I'll be using the Opal Kelly XEM6002 module for this build. Opal Kelly is an example of a good documentation provider: the XEM6002 User Manual (*http://assets00.opalkelly.com/ library/XEM6002-UM.pdf*) provides all the information we need in an easy-to-read format.

Looking at our block diagram of the XEM6002 module we can see that the PLL (phase lock loop) circuit block is clearly identified (green block). The PLL circuit is typically related to the master timing reference clock generation circuit (see Figure 2-48).

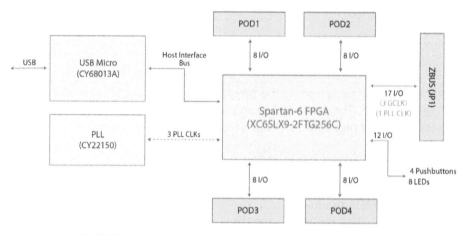

Figure 2-48 *XEM6002 module block diagram with PLL block at lower left*

We can see that the PLL generates three clocks that we can use in our designs. We'll need to go to the user manual for more information on these reference clocks. In the table of contents, we can see that there's a section on the PLL under "Introducing the XEM 6002." We find the following information about the PLL in this section:

Cypress CY22150 PLL

A multi-output, single-VCO PLL can provide up to five clocks, three to the FPGA and another two to the expansion connectors JP2 and JP3. The PLL is driven by a 48-MHz signal output from the USB microcontroller. The PLL can output clocks up to 150-MHz and is configured through the FrontPanel software interface or the FrontPanel API.

We now have a few more pieces of the puzzle. We know that the maximum frequency of the three FPGA input clocks is 150 Mhz and that we can configure their frequency from the

XEM6002 FrontPanel application that is running on the PC. That is a really nice feature! We also see that there are two additional master clocks that we have access to through the expansion connectors (also a nice feature).

We still don't have all the information we need, though, so we need to dig a little deeper. The next section in the table of contents of the user manual is titled "FPGA Pin Connections" (they're making this too easy!). In this section, we find a subsection called "PLL Connections." Bingo! That's what we're looking for. Going to that section, we find this description:

> The PLL contains six output pins, two of which are unconnected. The other four are labelled SYS_CLK1 through SYS_CLK4. SYS_CLK4 connects to JP1. The other three pins are connected directly to the FPGA. The table below illustrates the PLL connections.

PLL Pin	Clock Name	Connection
LCLK1	SYS_CLK1	FPGA - T8
LCLK2	SYS_CLK2	FPGA - K12
LCLK3	SYS_CLK3	FPGA - H4
LCLK4	SYS_CLK4	JP1 - Pin 1

Now we know that SYS_CLK1 is connected to FPGA physical pin T8. That's exactly what we were looking for. We won't need to go looking for the schematic at this point, which is just as well since Opal Kelly doesn't provide one for this module. If the manufacturer of the module you purchased doesn't provide you with such a well-thought-out user manual as Opal Kelly, then you will have to go to the schematic of the board (if available) and trace out the clock signals to find what FPGA pins they connect to.

Now we need to identify the connections for our RESET input signal and our two LGT output signals. We also need power and ground connections to complete our peripheral breadboard. These connections come from the general-purpose I/O (GPIO) connectors of the module. On the XEM6002, the GPIOs are called expansion connectors and labeled POD1 through POD4. We'll use POD1 for our connections to our LED breadboard. The user manual describes these connectors as follows:

> POD1 through POD4 connectors are 12-pin, dual-row, 0.1" female headers which have a pinout satisfying the Digilent Pmod specification. Two pins of each connector are connected to +3.3VDD (3.3V Power) and two pins are connected to DGND (Digital Ground). The remaining 8 pins are connected to the FPGA on banks 1 and 3.

POD1	Signal
1	M1
2	L1
3	K1
4	J1
5	DGND
6	+3.3VDD
7	E1
8	E2
9	F1
10	G1
11	DGND
12	+3.3VDD

We now have all the connection information we need. Using the information we gathered from the user manual, we can build the pin mapping table (shown in Table 2-2) for our design.

Table 2-2 *Pin mapping table*

Design signal	Direction	Board connection	FPGA pin
CLOCK_IN	Input	SYS_CLK1	T8
RESET	Input	POD1-1	M1
LGT1 (OUT_HIGH)	Output	POD1-2	L1
LGT2 (OUT_LOW)	Output	POD1-3	K1
Ground	N/A	POD1-5	N/A
Power	N?A	POD1-6	N/A

Creating the Constraints File

The next thing we need to do is create a constraints file to tell the ISE implementation tool where we want to make our physical connections. You can assign I/O signals to physical pins in your design through the PlanAhead tool in ISE. Using this tool, you can assign I/O locations, specify I/O banks, specify I/O standards, prohibit I/O locations, and create legal pin assignments using the built-in design rule checks (DRC) system. The constraints are saved to the user constraints file (UCF).

Follow these steps to open the PlanAhead tool:

1. In the Design panel, select the Implementation view.

2. In the Hierarchy pane, select the top module image or the associated UCF file.

3. In the Processes pane, expand User Constraints, and double-click "I/O Pin Planning (PlanAhead)." You can select Pre-Synthesis or Post-Synthesis in this case, because we already synthesized our design and don't care if we resynthesize.

The PlanAhead software opens with the PlanAhead environment loaded and extracts the top-level I/O port information from your associated source files. If a UCF file does not exist, an empty one is created for you.

Once you are in the editor all you need to do is type the FPGA pin assignments that are in your table into the Site configuration dialog box of your Scalar ports list, as shown in Figure 2-49.

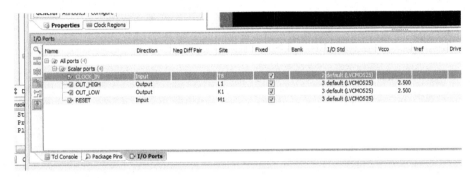

Figure 2-49 *Add pin assignments to site configuration*

After updating the constraints, you must save the PlanAhead project and exit the PlanAhead software. This updates the UCF files in the ISE project and updates the ISE project status accordingly. If you do not save and exit from the PlanAhead software, the ISE project and UCF files are not updated.

Heads Up

Make sure you update your constraints before you exit PlanAhead!

You can open the UCF file to look at what was generated by PlanAhead by just double-clicking on the UCF file in the Hierarchy pane. You should see something like Figure 2-50.

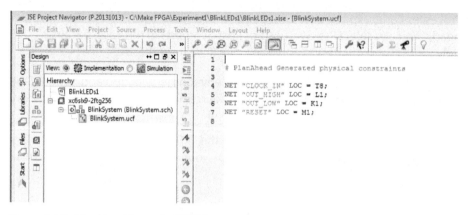

Figure 2-50 *PlanAhead Generated UCF file*

Did You Know?

Opal Kelly also provides a very nice application on its website to help with the UCF file generation called Pins. You can use the Pins application to generate your UCF file and then just place it in your ISE project directory. When using this method you must be sure you do not have any assignments in your UCF file that are not in your design, as this will cause ISE to generate errors when you try to reimplement the design.

We are now ready to generate our bit file.

Bit File Generation

Before you run the Generate Programming File process, click Implement Top Module (the green triangle) one more time, just to make sure everything is updated. The Generate Programming File process runs BitGen, the ISE bitstream generation program, which produces a bitstream (*.bit*) file for your FPGA device configuration. To initiate this process:

1. In the Design panel, select the Implementation view.

2. In the Hierarchy pane, select the top module image.

3. In the Processes pane, double-click Generate Programming File.

The programming file is saved in your project directory. After running this process, you are ready to program your target FPGA device.

Program Target Device

To program your target device, you will need to follow the instructions in your module's user manual. Unfortunately this is another one of those quirky issues with the ISE tool. Xilinx considers the programming protocol used in its iMPACT tool proprietary, and therefore only Xilinx modules can use this tool. Most other manufacturers of FPGA modules, like the ones we are using in this book, have come up with their own applications to program the Xilinx FPGAs on their modules. Opal Kelly makes this very easy through its FrontPanel application, which runs on your PC. You will need to consult the documentation of your particular module to find the exact programming methods and tools required.

Before we program our device, let's hook up our module to the breadboard we made earlier. First, get yourself oriented with the GPIO connector on your module. In this example, I'm using the POD1 connector of the XEM6002, so I connected my jumper wires to the POD connector pins following the table I constructed earlier (Table 2-2), as shown in Figure 2-51.

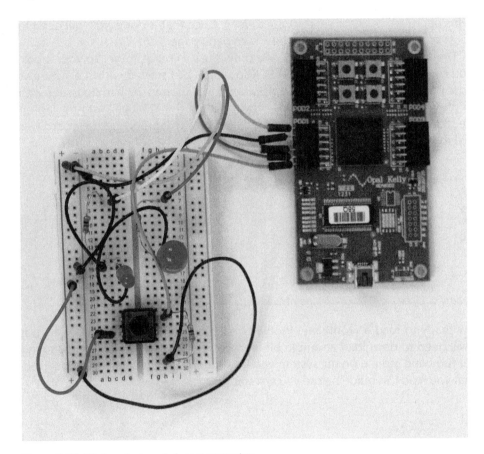

Figure 2-51 *Blink project ready for programming.*

Now all we need to do is hook up our USB cable to the module (or whatever connection method is required for power and loading the bit file), and we are ready to go.

Open Kelly Setup

Heads Up!

Please note that due to the rapidly changing world of FPGA development boards and development technology, the setup procedures described here are subject to change. The following procedures were accurate at the time of writing but may have changed since.

To get started with your Opal Kelly board, you will first need to create an account on the Opal Kelly Pins page (*https://www.opalkelly.com/pins/*). Once that is done, you will need to send a request to download the SDK via the Downloads page (*https://pins.opalkelly.com/downloads*). The approval response should take only a few minutes to get to your inbox, and then you can download the FrontPanel SDK for your OS. I started with the Windows x64 version, which I installed on a Lenovo laptop running Windows 7. I had no issues with the install. I plugged the USB cable that was provided with the XEM6002 into my laptop and into the XEM6002 module and opened FrontPanel on the laptop, and the board was there on the FrontPanel Welcome screen (Figure 2-52). So far, so good!

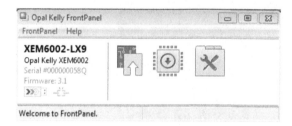

Figure 2-52 *You should see the XEM6002 on the Welcome screen*

You are now ready to load a FrontPanel Profile by clicking on the arrow icon. Well, not quite! You will need to download a sample bit file from the Opal Kelly web page. I found this out after fumbling around on my system and locating the Getting Started Guide. I recommend that you read this guide before you get started.

Heads Up!

*Be sure to read the **GettingStarted-USB.pdf** file in your **C:/Program Files/ Opal Kelly/FrontPanelUSB/Documentation** folder before you begin!*

Setup Test

You can locate sample bit files on the Downloads page of the Opal Kelly website (under the Support tab). You will need to download the ZIP file that contains the XEM6002 files (see Figure 2-53) and extract the files to a directory on your system.

FrontPanel Sample Bitfiles

FILE	DESCRIPTION	SIZE
20111228-FP40-XEM3001v2.zip	XEM3001 Sample Bitfiles	81.1 KB
20111007-FP40-XEM3005-1200.zip	XEM3005-1200 Sample Bitfiles	138 KB
20111007-FP40-XEM3010-1000.zip	XEM3010-1000 Sample Bitfiles	107 KB
20111007-FP40-XEM3010-1500P.zip	XEM3010-1500P Sample Bitfiles	117 KB
20111007-FP40-XEM3050.zip	XEM3050 Sample Bitfiles	143 KB
20111007-FP40-XEM5010-LX50.zip	XEM5010-LX50 Sample Bitfiles	347 KB
20111007-FP40-XEM5010-LX110.zip	XEM5010-LX110 Sample Bitfiles	551 KB
20111007-FP40-XEM6001.zip	XEM6001 Sample Bitfiles	123 KB
20121002-FP42-XEM6002-LX9.zip	XEM6002-LX9 Sample Bitfiles	181 KB
20111014-FP40-XEM6006-LX16.zip	XEM6006-LX16 Sample Bitfiles	229 KB

Figure 2-53 *Sample bit files on Opal Kelly site*

Now you are ready to test the setup by moving a sample file to the XEM6002 board. I followed the steps outlined in the section "An Introductory Project" of the Getting Started Guide and was able to successfully run my first FrontPanel example (Figure 2-54). I suggest that first-time users do the same.

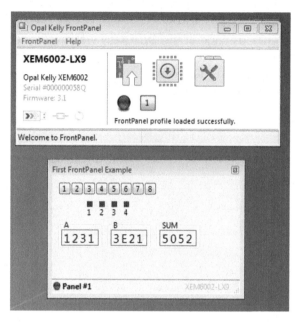

Figure 2-54 *Successful loading of "First" sample bit file and profile to XEM6002*

With Opal Kelly's FrontPanel successfully installed, and having determined that the board is working properly, all you need to do now is drag and drop your bit file onto the configuration icon, as shown in Figure 2-55.

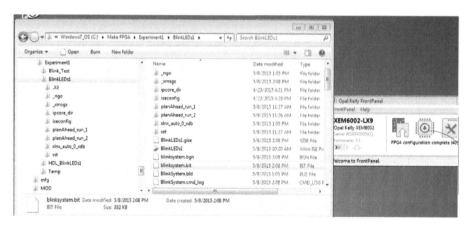

Figure 2-55 *Drag bit file onto configuration icon*

Once your programming is complete you should see the LEDs flashing, or just continuously on (this depends on your master clock reference frequency). Congratulations—you have completed your first FPGA design!

If you have no LEDs on at all, you will need to check your connections. If your LEDs are constantly on, then you will need to either slow down your master clock reference or use higher-order bits of your counter for OUT_HIGH and/or OUT_LOW.

Clock Frequency Experimentation

At this point you can go on to experiment with the frequency configuration of your module's master timing circuit or you can change the taps of your counter to be higher-order bits to get a lower-frequency clock. For example, instead of using bits 20 and 21, you can use 25 and 29. You will need to edit your schematic or Verilog source file and then reimplement and generate a new bit file to see the results of this on your LEDs.

Opal Kelly provides a great PLL configuration tool in FrontPanel that lets you configure the clock frequencies, as seen in Figure 2-56. You can do this while your design is running, and you don't need to reprogram the FPGA.

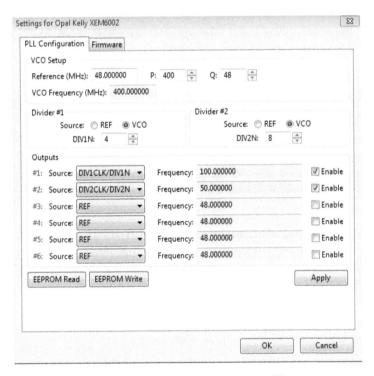

Figure 2-56 Opal Kelly FrontPanel PLL configuration utility

Takeaways

Here are some of the key takeaways from the exercises in this chapter:

- Binary counters make good frequency dividers.
- The human eye cannot detect flicker rates much above 100 Hz.
- Using a simple LED blink test and an FPGA frequency divider is a good vehicle to test the functionality of a new FPGA platform.
- Be sure to specify the correct FPGA device information for your FPGA platform when setting up your Xilinx ISE project.
- Don't forget to occasionally save your work as you create your project source file.
- Graphical design entry (schematics) and HDL (Verilog or VHDL) produce the same synthesis results for the most part.
- Schematics may be easier to follow at first, but it takes much more effort to create a graphical design than it does to use an HDL and a text editor.
- Flip-flops are digital logic circuits that can be in one of two states.
- The most common flip-flop is the D-flip-flop (DFF) or D register.
- Verilog requires a semicolon at the end of every code string.
- The RTL consists of two kinds of elements: registers and combinational logic. Registers (usually implemented as D flip-flops) synchronize the circuit's operation to the edges of the clock signal, and are the only elements in the circuit that have memory properties. Combinational logic performs all the logical functions in the circuit, and it typically consists of logic gates.
- Timing issues are the most challenging issues to debug in an FPGA design. For us, it's good to know that we will not be working with clock speeds that push the envelope of our FPGA technology. The ISE synthesis tools have come a long way over the years. The clock frequency can be set in the UCF file as a constraint, forcing the tool to attempt to meet the timing requirements.
- Doing a simple DOA simulation first, even for small designs, provides a good sanity check to see if the design works functionally at some level.
- Assigning physical I/O connections can be the most difficult part of the whole design process, because this is where you link the virtual world of your ISE design to the real world of the actual FPGA chip and your particular FPGA module's circuit board layout. Sometimes it can take a little bit of detective work to pull all the pieces together; some FPGA module vendors do a better job of providing documentation than others.

- Create a constraints file (UCF) to tell the ISE implementation tool where you want to make your physical connections to the FPGA device.

- To program your target device you will need to follow your module's user manual's instructions. Xilinx considers the programming protocol used in its iMPACT tool proprietary, so only Xilinx modules can use this tool. Most other manufacturers of FPGA modules have come up with their own applications to program the Xilinx FPGAs on their modules.

That's Refreshing | 3

Concurrent HDL Code Execution with Seven-Segment Displays

In this chapter, I'll be getting you familiar with the fundamental concept of HDL concurrency in FPGA design. The whole idea of HDL concurrency may be somewhat strange to you at first, especially if you are coming from a software programming background or have done some projects with an embedded microcontroller like an Arduino and created sketches in the C programming language for it.

The first thing you need to wrap your head around is that when you are coding in HDL, you are not writing a software program; rather, you are describing digital hardware logic functionality. With concurrency, there are no sequential steps of code execution like: "first do this, then do this, then do that." There really is only one instant in time, and that is the clock tick. Remember our RTL model of the binary counter in the last chapter: it had the current-state vector stored in flip-flops. It used this vector as feedback and combined it with other inputs (RESET) to generate the next-state vector. We saw that when the clock ticked the next state, which was the count +1, became the current state, which was the current count.

Think of the HDL you are using as more like describing a block diagram than a flow chart. In a block diagram representation, all the functional blocks can be interdependent and in operation simultaneously. In other words, the block operations are concurrent processes. On the other hand, a flow chart typically represents a sequential process flow.

To help you understand this concept of concurrency, we'll be building a 7-segment display stopwatch circuit that measures elapsed time in seconds with 100 millisecond (1/10 of a second) resolution. We'll also be looking at configuring and instantiating a digital clock manager (DCM) core in our design.

Stopwatch Concept

We will be building on the binary counter design that we completed in the last chapter. Our stopwatch is another relatively simple design that can be described with a high-level block diagram shown in Figure 3-1.

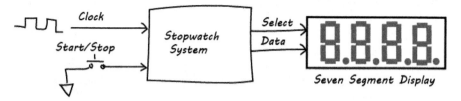

Figure 3-1 *Stopwatch system*

The input clock is the master timing reference, which will provide the timebase for the seconds counter. Pushbuttons will provide us with the physical mechanism to start, stop, and clear the elapsed time count. When a button is pressed, elapsed time is displayed with millisecond resolution; when a button is released, the elapsed time will hold and the current elapsed time count will be displayed. When another button is pressed, the display will reset to all zeros.

How It Works

To understand the theory of operation for this design, we will need to review how a 7-segment display works. Seven-segment displays are used to display decimal digits and some letters. Each digit display element has seven inputs, one for each of its segments. The data signals control which segments are highlighted. By selecting which segments are highlighted, several characters can be displayed. For example, every decimal digit can be displayed as illustrated in Figure 3-2.

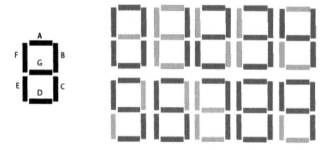

Figure 3-2 *7-segment display decimal digits*

The 7-segment display is made up of seven individual LED elements, or diodes. The diode is a one-way electronic valve that allows current to flow through it when its anode is positive with respect to its cathode. The diode approximates an open switch when it is in the reverse-bias state and a closed switch when forward-biased (Figure 3-3). The LED emits light when it is in the forward-biased state.

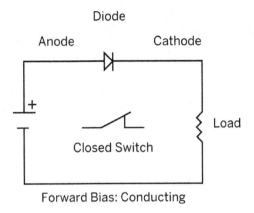

Forward Bias: Conducting

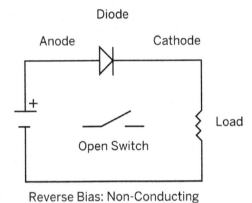

Reverse Bias: Non-Conducting

Figure 3-3 *Diode biasing*

The seven diodes for each character can be wired in a common cathode or common anode configuration. We'll be using a 7-segment, LED display (HS410561K-32 common anode). This is a 12-pin device. You can see from the schematic representation of this device (Figure 3-4) that all the anodes of each digit are wired together (common anode). Common cathode configuration is the opposite, with all the cathodes wired together (see Figure 3-5).

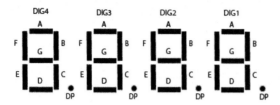

Figure 3-4 *Four-digit 7-segment display*

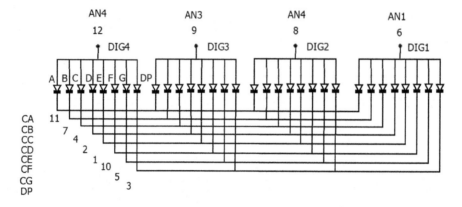

Figure 3-5 *Pin assignments*

Heads Up

You need to know if your 7-segment display is a common cathode or common anode configuration. This will determine how you connect it to your FPGA and drive it to the "on" state.

Design

We will be using Verilog HDL once again for our FPGA design, and we will also need to build another small breadboard circuit that will act as our peripheral device. This breadboard will contain our 7-segment display, which will connect to the FPGA module with jumper wires. We'll also need to add two push buttons for the "clear" and "hold" functions. Many FPGA boards include 7-segment displays. You will need to check your user manual or module schematic to understand the pin out of the display that is on your module if you decide to use it.

Peripheral Breadboard

I used a half-size breadboard to wire up my 7-segment display following the schematic in Figure 3-6.

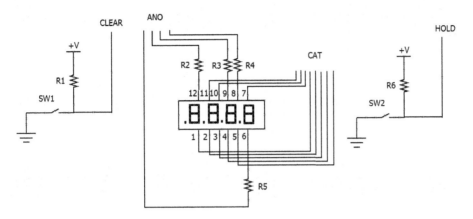

Figure 3-6 *Display breadboard schematic*

BoM

- 1 × HS410561K-32 4-digit 7-segment display
- R1, R2, R3, R4, R5 = 330 - 1K Ohm (any value of resistor in this range will work)
- SW1, SW2 = 2 × push button switches
- 1 × small (1/2 size) breadboard
- 14 × male-to-male jumper wires

Since the four digits of our display share the same data bus (CAT bus), the multiplexing of the data to the display is done by our FPGA circuit. To drive all digits with the same value, we drive ANO 1...4 high. To drive different values we change the output on the CAT bus with the appropriate ANO signal. Since this transition can happen at a high frequency, it gives the appearance that all digits are "on" simultaneously, when they are actually being turned on one at a time and cycled rapidly. Remember our discussion of flicker rate and human eye detection from the last chapter? In this case we are doing the opposite by speeding up the refresh rate well beyond the 100 Hz rate the human eye can detect. In this way we are actually cycling through the four characters one at a time, but at such a fast rate that our eyes only see the four digits as "on" simultaneously. This is the same way a computer monitor or TV screen works with refresh rates. All those pixels in your monitor are changing very rapidly—so fast that your eye does not detect the changes but sees constant images.

We'll use two push buttons to control the start/stop functions. SW2 acts like a stopwatch function. While you are holding down the button, the stopwatch is running. When you release the button, the count freezes and the elapsed time is displayed.

FPGA Circuit

To start our design, we will follow these steps:

1. Open a new ISE project and name it *SevenSegmentDisplay1*. Don't forget to select HDL as the top-level source type.

2. Click Next and in the Project Settings dialog box, select Verilog as your preferred language.

3. Click Next and open a new source file from the Project menu in the top toolbar.

4. Select Verilog Module from the Select Source Type dialog box and name your file *DSPDRV*.

5. Click Next and define your module's ports, as illustrated in Example 3-1.

Example 3-1 *Port definitions*

```
module SW7SegDisplayDriver(
CAT , //Output to cathodes
ANO , //Output to anodes
CLK , //Clock input
CLEAR, //Reset input
HOLD, //Hold input
    );
//------Output Ports-------
output[7:0]CAT;
output[3:0]ANO;

//------Input Ports--------
input CLK, CLEAR, HOLD;
```

The design of the 7-segment display driver circuit is relatively straightforward. We can see from the block diagram in Figure 3-7 that the timing reference (clock) from our platform master clock comes into the first block on the left, which is labeled "Clock Generation." This block is the DCM, which provides the system clock (sysclk, labeled "slow_clk" in our design).

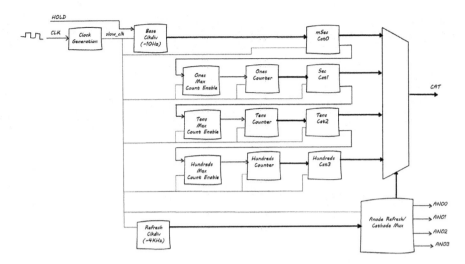

Figure 3-7 *7-segment display driver FPGA design*

The "Base Clkdiv" block provides us with our 100 ms (1/10 of a second) time base. The output of this block will be used to generate the cathode codes for the millisecond digit position. You can see that there are four blocks that generate cathode codes: one block for each digit position. These four blocks are connected to a four-to-one multiplexer, which takes the four independent digit cathode codes and muxes them onto our single cathode (CAT) output bus.

The multiplexer's output selection is controlled by the "Anode Refresh/Cathode Mux" block, whose timing is controlled by the "Refresh Clkdiv" block. We need to run our refresh rate much faster than our count. I chose to run it at around 1 KHz. The refresh circuit is cycling through the cathode output for each digit approximately every millisecond; at the same time, the anode is selected for that digit. We just see the four digits displayed; we don't see each digit being updated sequentially because it is happening much too fast for our eyes to detect.

The 100 ms (1/10 of a second) cathode code block also generates an output signal that indicates the terminal count of the 100 ms counter (1,000 ms). In this case the terminal count is 10. This signal is then used to generate an enable strobe (one clock cycle pulse) to the ones counter in the "Ones Max Count Enable" block. For every 1,000 ms count, we will increment our ones counter. This then feeds our tens position, and the tens position feeds our hundreds position.

The HOLD input, when active, causes our circuit to count by enabling the "Base Clkdiv" block. When HOLD is not active, the count will not advance—in other words, the current count will equal the next count. Not shown is our RESET input. When RESET is active, all counts are cleared (reset to zero).

Generating Enable Strobes

All digital logic designers have their bag of tricks that they use again and again. Generating a single clock pulse strobe out of a very long multiclock-period signal is one of them.

You can do this easily with just a couple of lines of HDL code:

```
always @(posedge clk)
begin
    reg_maxcnt <= MAXCNT;
    ENABLE <= (maxcnt & !reg_maxcnt);
end
```

The schematic is shown in Figure 3-8.

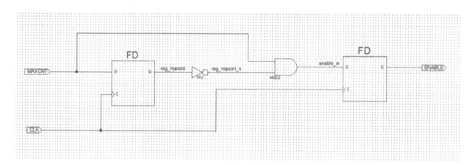

Figure 3-8 *Schematic*

Looking at a simulation of this, you will see a nice one-clock pulse at the rising edge of each MAXCNT signal (Figure 3-9). This is a very handy circuit to have in your back pocket.

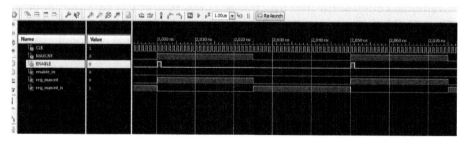

Figure 3-9 *Simulation*

Digital Clock Manager (DCM)

We learned in the last chapter the basic concept that if we want to divide a clock by n, we simply have a counter count to $n/2$. So, for example, if we wanted to divide a clock by 9, we'd use a counter to count to 4. We can easily see the limitations of this method in that we can't scale by odd numbers as we'll end up with a noninteger value. We also can't use this method to generate a clock that is at a higher frequency than our input clock. However, if we use the DCM function that Xilinx provides from within the ISE tool and instantiate this DCM module in our FPGA design, we can get most clock frequencies that we need easily.

For our 7-segment display driver design we'll be using a 100 MHz input clock master. Now we need to divide the input clock down to a frequency for our 100 ms time base counter. We also need a much higher-speed clock to generate our refresh timing, so something on the order of 1 KHz (which is about 10 times faster than our eyes can detect, so there should be no flicker) will work fine. Taking all this into account, I decided to use the DCM to synthesize an 11.534 MHz output clock from the 100 MHz input clock. We'll use this clock and a 24-bit counter to generate my 10 Hz time base. I didn't pick 11.534 MHz out of the air, if you are wondering. You will see that if we use the last 4 bits of our 24-bit counter (bits 23:20) to count to 10, that is binary 1010 (we are only counting to 10 because we are displaying the base 10 number system). Then we have the binary number:

        1010 1111 1111 1111 1111 1111

This binary number represents our full 24-bit terminal count. If we convert this number to decimal, it equals 11,534,335, or 11.534×10^6. So, if we want to have our counter take 1 second to count in 24-bit times, we will need to clock it at 11.534 MHz, or once every 86.69 nanoseconds for 11,534,335 ticks of the clock.

Let's try another example. Let's consider saving some FPGA register resources in constructing the time base counter. You can do this by using a lower-frequency DCM output clock and a smaller counter. If you do this, then you will need to move the 4 bits used to count to 10 to the right in the bit ordering to align your terminal count bit position with your frequency. For example, you will use a 20-bit counter instead of a 24-bit counter, then you will have 1010 1111 1111 1111, which is decimal 45,054. This will require a system clock of 45.054 KHz or 0.0450 MHz as your DCM output clock frequency. Note that you may need to adjust your refresh counter as you drop the system clock down. I used a 13-bit counter to generate the refresh time base, which is about 1 KHz using the 11.538 MHz clock.

You may be wondering why we need a counter at all. The answer is simple: it's because the DCM can't synthesize down to a 10 MHz clock from a 100 MHz source.

Heads Up!

Remember that all my calculations were based on the 100 MHz input clock from my platform's master timing generator. You will need to base your design on your input clock frequency if it is not 100 MHz!

Now to generate our DCM, we need to use the Xilinx's core generator in our ISE tool. To do this:

- In your 7-segment display ISE project, right-click the device in the hierarchy and click New Source.

- Next, click IP (CORE Generator & Architecture Wizard), then give it a filename and click Next. I used *DCM100to8_3886*.

- Select FPGA Features and Design→Clocking→Clocking Wizard, then click Next and Finish. The Clocking Wizard will open, as seen in Figure 3-10.

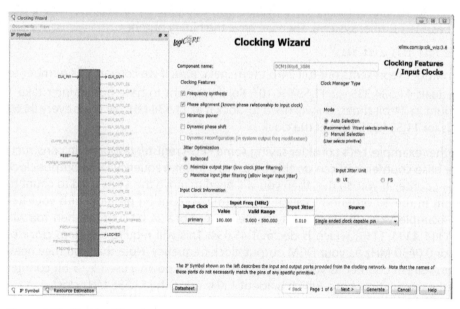

Figure 3-10 *Clocking Wizard*

- Set the input frequency to 100 MHz (if you're using the Opal Kelly board; otherwise set the frequency appropriately for your own hardware), then click Next.

- Set the output frequency to 11.534 MHz, then click Next.

- Deselect RESET and LOCK, then accept the defaults for the rest of the wizard.

- Click Generate This to create a new core that implements a DCM that divides the input clock frequency.

DCM Synthesized Clock Accuracy

When I was configuring my DCM, the closest output clock frequency to the desired 11.534 MHz that the DCM could generate using the 100 MHz input clock was 11.538 MHz. You can see that this is a slightly faster clock than the calculated required value of 11.534 MHz. This will result in an accuracy error for our time base of about 0.035%, which is on the fast side. When dealing with digital frequency synthesizers it is common to only get close to the calculated value required for some combinations of input and output clocks. In our case, for this exercise, the accuracy is not critical and the error really has no impact. But, for example, if you are using the DCM to generate a serial communication baud rate clock, then the error and accuracy of the baud clock will be critical in meeting the particular serial protocol specification and for avoiding bit errors in the transmission and/or reception of data.

You can then instantiate this module core in your Verilog by simply using the following code:

```
/Digital Clock Manager Module - Generated by ISE
DCM_INCLKtoSYSCLK timebase_dcm ( .CLK_IN1(CLK), // Clock input (100MHz)
                .CLK_OUT1(sysclk) ); // Clock out (11.538MHz)
```

Verilog Modules

*Modules are the building blocks of Verilog designs. Modules can be instantiated from within other modules—this is hierarchical design. When a module is instantiated, connections to the ports of the module must be specified. There are two ways to make port connections. One is **connection by name**, in which variables connected to each of the module's inputs or outputs are specified in a set of parentheses following the name of the port. This is the method used in the DCM module example. In this method order of connections is not significant. The second method is called **ordered connection**. With this method, the order of the ports must match the order in which they appear in the instantiated module. When ports are connected by name, it is illegal to leave any ports unconnected, but unconnected ports may occur when ports are connected by order. The ordered connection method is prone to problems as the user makes changes to a module and then has to connect it in the proper order.*

The Xilinx FPGA digital clock manager core is a very versatile and complex piece of IP. It can be used to implement a delay-locked loop, a digital frequency synthesizer, a digital phase shifter, or a digital spread spectrum. Most of these topics are beyond the scope of this book, and I will not be covering them. For our purposes, we will be focusing on using it as a simple digital frequency synthesizer. If you would like to explore more of the DCM's functions and capabilities, I recommend that you take a look at these documents:

- The Xilinx, Digital Clock Manager (DCM) Module Product Specification (*http:// bit.ly/1JWmNGX*)

- The Xilinx Spartan-6 FPGA Clocking Resources User Guide (*http://bit.ly/1JWmMTE*)

Verilog Code and Concurrency

This section provides you with my complete Verilog code for our FPGA, 7-segment display driver. I purposely wrote this code in a style that easily highlights the concept of concurrency. The code follows the block diagram I presented earlier in the chapter (Figure 3-7), so it should be easy for you to dissect it.

In HDL, a structured procedure statement provides the means to specify concurrency. The *always* block is the Verilog construct for running concurrently. You can have one or multiple procedural assignment statements within an *always* block. The Verilog syntax for the *always* block is shown in Example 3-3.

Example 3-3 *Syntax of always block*

```
always @(sensitivity_list)
begin
//one or more procedural assignment
//statements
end
```

Assignments inside the block must be enclosed within the begin and end statements, which encapsulate the procedural assignment statements that can run sequentially. The sensitivity list decides when the always block is entered. Typically this is the system clock's rising edge (posedge) or falling edge (negedge). The nonblocking assignment statement (<=) provides the evaluation of the scheduled event and models the hardware behavior, which is illustrated in Example 3-4.

Example 3-4 *Nonblocking assignment statement*

```
LeftSide <= RightSide

//Evaluate 'RightSide', schedule the value to 'LeftSide' and proceed.
//The values are assigned at the end of current simulation time
```

```
//signified by the clock tick.
//Remember that the execution order is independent of statement order!
```

When you study the full code of the 7-segment stopwatch driver (listed in Example 3-5), you will see many always blocks. Actually, there is an always block description for every block represented in our block diagram. These always blocks are modeling the hardware behavior of all the logic in the circuit. These blocks are evaluated at the positive edge of the system clock, or in other words, at the same instant in time. So, the code is not sequential in the sense of, for example, the base clock circuit runs before the Refresh Clock circuit, even though you can plainly see in the code that one is before the other. They all are evaluated at the same instant in time. This type of concurrency is not explicit in typical computer programming languages like C, and you'll need to get accustomed to thinking about and reading HDL code in this new way.

 FYI

*You can find this code on this book's GitHub page (**https://github.com/ tritechpw/Make-FPGA**).*

Example 3-5 *Full Verilog code for 7-segment display stopwatch driver*

```
`timescale 1ns / 1ps
///////////////////////////////////////////////////////////////////////

// Module Name:    SW7SegDisplayDriver

///////////////////////////////////////////////////////////////////////
module SW7SegDisplayDriver(
CAT,        //Output to cathodes
ANO,        //Output to anodes
CLK,        //Clock input
CLEAR,      //Reset input
HOLD,       //Hold input

    );
//------Output Ports-------
output[7:0]CAT;
output[3:0]ANO;

//------Input Ports--------
input CLK, CLEAR, HOLD;

//-----Internal Variables-----
reg[23:0]clkdiv;
reg[23:0]cathode;
```

```verilog
reg[7:0]cat0;
reg[7:0]cat1;
reg[7:0]cat2;
reg[7:0]cat3;

reg[13:0]refclk;
reg[3:0]anode;

reg[3:0]ones;
reg[3:0]tens;
reg[3:0]hundreds;

reg onescnt;
reg tenscnt;
reg hundredscnt;

reg onesmaxcnt_en;
reg tensmaxcnt_en;
reg hundredsmaxcnt_en;

reg onesmaxcnt;
reg tensmaxcnt;
reg hundredsmaxcnt;

//---- internal signals
wire hold_in;
wire reset;
wire sysclk;

//Digital Clock Manager Module - Generated by ISE
SW7SegDisplayDriver timebase_dcm ( .CLK_IN1(CLK), // Clock input (100 MHz)
                                   .CLK_OUT1(sysclk) ); // Clock out (11.534 MHz)


//------Code Starts Here------

assign CAT = cathode;
assign ANO = anode;
assign reset = !CLEAR;
assign hold_in = HOLD;

//------- Base clkdiv - 100 ms
always @(posedge sysclk)
 if (reset) begin
   clkdiv <= 24'b0 ;
 end
 else /// hold
 if (hold_in) begin
   clkdiv <= clkdiv;
   end /// end hold
 else
 if (clkdiv [23:20] == 4'b1010) begin // Terminal count
  clkdiv <= 24'b0;
 end
```

```
  else begin
   clkdiv <= clkdiv + 1;
  end

//------- Refresh clkdiv 1 KHz (1 ms)
always @(posedge sysclk)
 if (reset) begin
  refclk <= 25'b0;
 end
 else
 begin
  refclk <= refclk + 1;
 end

//------- ones max count enable
always @(posedge sysclk)
 if (reset) begin
  onescnt <= 1'b0;
 end
 else
 begin
   onescnt <= onesmaxcnt;
   onesmaxcnt_en <= (onesmaxcnt & !onescnt);
 end

//------- tens max count enable
always @(posedge sysclk)
 if (reset) begin
  tenscnt <= 1'b0;
 end
 else
 begin
  tenscnt <= tensmaxcnt;
  tensmaxcnt_en <= (tensmaxcnt & !tenscnt);
 end

//------- hundreds max count enable
always @(posedge sysclk)
 if (reset) begin
  hundredscnt <= 1'b0;
 end
 else
 begin
  hundredscnt <= hundredsmaxcnt;
  hundredsmaxcnt_en <= (hundredsmaxcnt & !hundredscnt);
 end

//------- ones count
always @(posedge sysclk)
 if (reset) begin
  ones <= 4'b0000;
 end
 else
 if (onesmaxcnt_en == 1'b1) begin
```

```
    ones <= ones + 1;
   end
   else if (ones[3:0] == 4'b1010) begin
     ones <= 4'b0000;
   end

//------- tens count
always @(posedge sysclk)
 if (reset) begin
  tens <= 4'b0000;
 end
 else
 if (tensmaxcnt_en == 1'b1) begin
   tens <= tens + 1;
  end
  else if (tens[3:0] == 4'b1010) begin
    tens <= 4'b0000;
   end

//------- hundreds count
always @(posedge sysclk)
 if (reset) begin
  hundreds <= 4'b0000;
 end
 else
 if (hundredsmaxcnt_en == 1'b1) begin
  hundreds <= hundreds + 1;
 end
 else
 if (hundreds[3:0] == 4'b1010) begin
  hundreds <= 4'b0000;
 end

// ----- ms base count ------
always@(posedge sysclk)
 if (reset) begin
  cat0 <= 8'b11000000;
  onesmaxcnt <= 1'b0;
 end
 else
 if(clkdiv [23:20] == 4'b0000) begin
  cat0 <= 8'b11000000;//0
  onesmaxcnt <= 1'b0;
 end
 else
 if (clkdiv [23:20] == 4'b0001) begin
  cat0 <= 8'b11111001;//1
 end
 else
 if (clkdiv [23:20] == 4'b0010) begin
   cat0 <= 8'b10100100;//2
 end
 else
 if (clkdiv [23:20] == 4'b0011) begin
```

```
  cat0 <= 8'b10110000;//3
end
else
if (clkdiv [23:20] == 4'b0100) begin
 cat0 <= 8'b10011001;//4
end
else
if (clkdiv [23:20] == 4'b0101) begin
 cat0 <= 8'b10010010;//5
end
else
if (clkdiv [23:20] == 4'b0110) begin
 cat0 <= 8'b10000010;//6
end
else
if (clkdiv [23:20] == 4'b0111) begin
 cat0 <= 8'b11111000;//7
end
else
if (clkdiv [23:20] == 4'b1000) begin
 cat0 <= 8'b10000000;//8
end
else
if (clkdiv [23:20] == 4'b1001) begin
 cat0 <= 8'b10011000;//9
end
else
if (clkdiv [23:20] == 4'b1010) begin
  onesmaxcnt <= 1'b1;
end

// ----- Sec count ------
always@(posedge sysclk)
 if (reset) begin
  cat1 <= 8'b11000000;
  tensmaxcnt <= 1'b0;
 end
 else
 if(ones [3:0] == 4'b0000) begin
  cat1 <= 8'b01000000;//0
  tensmaxcnt <= 1'b0;
 end
 else
 if (ones [3:0] == 4'b0001) begin
  cat1 <= 8'b01111001;//1
 end
 else
 if (ones [3:0] == 4'b0010) begin
  cat1 <= 8'b00100100;//2
 end
 else
 if (ones [3:0] == 4'b0011) begin
  cat1 <= 8'b00110000;//3
 end
```

```verilog
    else
    if (ones [3:0] == 4'b0100) begin
     cat1 <= 8'b00011001;//4
    end
    else
    if (ones [3:0] == 4'b0101) begin
     cat1 <= 8'b00010010;//5
    end
    else
    if (ones [3:0] == 4'b0110) begin
     cat1 <= 8'b00000010;//6
    end
    else
    if (ones [3:0] == 4'b0111) begin
     cat1 <= 8'b01111000;//7
    end
    else
    if (ones [3:0] == 4'b1000) begin
     cat1 <= 8'b00000000;//8
    end
    else
    if (ones [3:0] == 4'b1001) begin
     cat1 <= 8'b00011000;//9
    end
    else
    if (ones [3:0] == 4'b1010) begin
     tensmaxcnt <= 1'b1;
    end

// ----- tens count ------
always@(posedge sysclk)
 if (reset) begin
  cat2 <= 8'b11000000;
  hundredsmaxcnt <= 1'b0;
 end
 else
 if(tens [3:0] == 4'b0000) begin
  cat2 <= 8'b11000000;//0
  hundredsmaxcnt <= 1'b0;
 end
 else
 if (tens [3:0] == 4'b0001) begin
  cat2 <= 8'b11111001;//1
 end
 else
 if (tens [3:0] == 4'b0010) begin
  cat2 <= 8'b10100100;//2
 end
 else
 if (tens [3:0] == 4'b0011) begin
  cat2 <= 8'b10110000;//3
 end
 else
 if (tens [3:0] == 4'b0100) begin
```

```verilog
    cat2 <= 8'b10011001;//4
  end
 else
 if (tens [3:0] == 4'b0101) begin
  cat2 <= 8'b10010010;//5
 end
 else
 if (tens [3:0] == 4'b0110) begin
  cat2 <= 8'b10000010;//6
 end
 else
 if (tens [3:0] == 4'b0111) begin
  cat2 <= 8'b11111000;//7
 end
 else
 if (tens [3:0] == 4'b1000) begin
  cat2 <= 8'b10000000;//8
 end
 else
 if (tens [3:0] == 4'b1001) begin
  cat2 <= 8'b10011000;//9
 end
 else
 if (tens [3:0] == 4'b1010) begin
  hundredsmaxcnt <= 1'b1;
 end

// ----- hundreds count ------
always@(posedge sysclk)
 if (reset) begin
  cat3 <= 8'b11000000;
 end
 else
 if(hundreds [3:0] == 4'b0000) begin
  cat3 <= 8'b11000000;//0
 end
 else
 if (hundreds [3:0] == 4'b0001) begin
  cat3 <= 8'b11111001;//1
 end
 else
 if (hundreds [3:0] == 4'b0010) begin
  cat3 <= 8'b10100100;//2
 end
 else
 if (hundreds [3:0] == 4'b0011) begin
  cat3 <= 8'b10110000;//3
 end
 else
 if (hundreds [3:0] == 4'b0100) begin
  cat3 <= 8'b10011001;//4
 end
 else
 if (hundreds [3:0] == 4'b0101) begin
```

```
    cat3 <= 8'b10010010;//5
    end
    else
    if (hundreds [3:0] == 4'b0110) begin
     cat3 <= 8'b10000010;//6
    end
    else
    if (hundreds [3:0] == 4'b0111) begin
     cat3 <= 8'b11111000;//7
    end
    else
    if (hundreds [3:0] == 4'b1000) begin
     cat3 <= 8'b10000000;//8
    end
    else
    if (hundreds [3:0] == 4'b1001) begin
     cat3 <= 8'b10011000;//9
    end

// ----- Anode Refresh/Cathode Mux ------
    always@(posedge sysclk)
     if (reset) begin
      anode <= 4'B0000;
      cathode <= 8'b11000000;
     end
     else
     if(refclk [12:11] == 2'b00) begin
      anode <=4'b0001;
      cathode <= cat0;
     end
     else
     if(refclk [12:11] == 2'b01) begin
      anode <=4'b0010;
      cathode <= cat1;
     end
     else
     if(refclk [12:11] == 2'b10) begin
      anode <=4'b0100;
      cathode <= cat2;
     end
     else
     if(refclk [12:11] == 2'b11) begin
      anode <=4'b1000;
      cathode <= cat3;
     end

    endmodule
```

You should be able to copy this into a new source file in your project and implement it. We'll do a quick simulation next to make sure everything is working as expected. (Note that there are other ways this mux could be coded; for example, as a case statement.)

Simulation

We'll be using the ISE waveform viewer again to do a quick DOA test of our design. To do this:

1. Click View Simulation in your design console.

2. Double-click Simulate Behavioral Model.

3. Once the Sim window is open, force the CLK signal to have a leading edge value of 1, a trailing edge value of 0, and a period of 10000 (10 ns or 100 MHz).

4. Force the CLEAR signal to a constant 0.

5. Force the HOLD signal to a constant 1.

6. Click Run for the time specified on the toolbar, leaving this at the default 1.00 µs.

7. Force the CLEAR signal to constant 1.

8. Force the HOLD signal to constant 0.

9. Click Run for the time specified on the toolbar.

10. Click Zoom to Full View.

You should see something like the waveform view in Figure 3-11. There should be no red signals (indicating an unknown state) after CLEAR goes high. This shows that the system is coming out of CLEAR in a stable state.

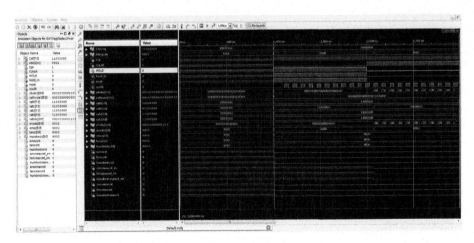

Figure 3-11 *DOA simulation*

Now zoom in and check that your DCM is functioning correctly. You can do this by adding another mark—right-click your waveform view window and selecting Markers→Add Markers (see Figure 3-12).

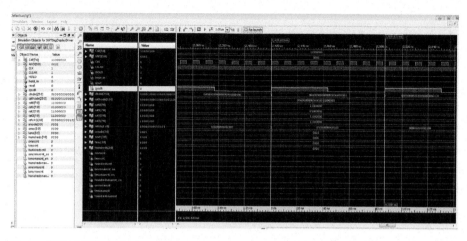

Figure 3-12 *DCM sysclk*

We can see that the sysclk period is around 86.800 ns, which is close to our 86.670 ns design target. Zooming out to our full extent again, we can also see that our clkdiv[23:0] is counting along with our refclk[13:0]. Things look healthy. Doing a full one second of simulation to test every function of our code would take hours, if not days. For this reason, we will move on to building our design and checking it in real-time on the bench.

We could alter the test parameters in the simulation to reduce the frequency of the input clock to speed up the simulation results, but even then it would still take a very long time to simulate the full one second of operation. This is a classic problem with simulation validation, where you need to trade simulation runtime for validation coverage. One way to shorten the runtime is to force signals, like the enables, to be active to see if logic behaves correctly for that condition.

Build

We can now do our final build and generate our bit file for loading onto our FPGA platform. Like in the last chapter, you will need to assign the physical I/O pins of our FPGA to our design.

Assigning Physical I/O

The process for assigning the GPIO pins is the same as last time (refer to "Build"); you will need to know your module's GPIO to FPGA pin mapping in order to do this. I'll be using the Open Kelly module again.

My constraints file is shown in Figure 3-13.

Now we just need to connect our breadboard, upload our bit file like before, and check out the operation of our design!

```
 1   #Start of Constraints
 2
 3   # I/O Pin Assignments
 4
 5   # Anode Bus                      7-Seg #Pin     - PMod - Pin
 6   NET "ANO<0>"   LOC = "M1"  ; #12           -P1-1
 7   NET "ANO<1>"   LOC = "L1"  ; #9            -P1-2
 8   NET "ANO<2>"   LOC = "K1"  ; #8            -P1-3
 9   NET "ANO<3>"   LOC = "J1"  ; #6            -P1-4
10
11   # Cathod Bus
12   NET "CAT<0>"   LOC = "E1"  ; #11           -P1-7
13   NET "CAT<1>"   LOC = "E2"  ; #7            -P1-8
14   NET "CAT<2>"   LOC = "F1"  ; #4            -P1-9
15   NET "CAT<3>"   LOC = "G1"  ; #2            -P1-10
16   NET "CAT<4>"   LOC = "A2"  ; #1            -P2-1
17   NET "CAT<5>"   LOC = "B2"  ; #10           -P2-2
18   NET "CAT<6>"   LOC = "B3"  ; #5            -P2-3
19   NET "CAT<7>"   LOC = "A3"  ; #3            -P2-4
20
21   # Mater Clock
22   NET "CLK"  LOC = "T8"  ;      #N/A         -NA-NA
23
24   # Clear
25   NET "CLEAR"  LOC = "F15"  ; #N/A           -P3-1
26
27   #Hold
28   NET "HOLD"  LOC = "G16"  ; #N/A           |-P3 -2
29
30   #End of Constraints
```

Figure 3-13 *UCF file for 7-segment display driver using Open Kelly module*

Takeaways

Here are the key takeaways from the exercises in this chapter:

- Real hardware operates concurrently.

- Concurrency is what differentiates HDL from other programming languages, which are sequential. Concurrency is not explicit in programming languages like C.

- The DCM core provides an easy way to perform frequency synthesis of clocks.

- When dealing with digital frequency synthesizers, it is common to only get close to the calculated value required for some combinations of input and output clocks.

- Modules are the building blocks of Verilog designs. Modules can be instantiated from within other modules.

- A classic problem with simulation validation is where you need to trade simulation runtime for validation coverage.

- The 7-segment display is made up of seven individual LED elements that are wired in common anode or cathode configurations.

Testing 1, 2, 3, 4 — 4

The HDL Test Bench

In the world of SoC design, the nature and methods of circuit verification is a topic that could fill volumes, let alone this entire book. We have already seen in the previous chapters how a simple simulation can be used to check the overall health of our designs. But what happens when we move beyond a simple counter design that we've created to using complex, off-the-shelf IP, like a communication controller or microcontroller? How do we test something like this, and how can we use simulation to help us understand its operation? Fortunately for us, a lot of clever engineers have spent a lot of time coming up with innovative ways to accomplish these goals. The *test bench* is one of these innovations, and it has become the primary vehicle used in industry for SoC verification. Designers sometimes also refer to a test bench as a *test fixture* (the two terms can be considered synonymous).

In this chapter, I will provide a basic understanding of what an HDL test bench is and what it does. More importantly, I will be walking you through the process of taking an existing IP block that includes a test bench and becoming familiar enough with it to be able to use it in a design project.

The Test Bench

In the simplest sense, a test bench is a virtual testing environment used to verify that a design does everything it's supposed to do and doesn't do anything it's not supposed to do. There are different styles of writing test benches, and in industry these styles are called *methodologies*. Over the years, test bench methodologies have evolved, with some of the more popular ones currently revolving around the Open Verification Methodology (OVM), a verification methodology developed jointly in 2007 by Cadence Design Systems and Mentor Graphics. OVM is a fairly advanced verification topic—you can even find university-level courses on OVM and its derivatives. For our purposes we'll be focusing on the basics of test bench construction by working with a Verilog test bench example. We'll be looking at a handful of simulation techniques that can be used on many of the digital applications we are experimenting with.

Test Bench Anatomy

Before we begin, let's review some of the basic components and terminology of a typical test bench. A test bench applies stimuli (inputs) to the unit under test (UUT), also referred to as the device under test (DUT). It monitors the outputs, providing status and error reporting in a readable and user-friendly format.

The test bench wrapper (Figure 4-1) typically is responsible for displaying values in a terminal window, generating results in a waveform viewer, and checking the correctness of the functional operation of the DUT. Typically there are no ports for the test bench itself. In other words, nothing is connected to it. The preferred method of instantiation of the DUT into the test bench is by name association, where each DUT port signal is explicitly associated with a test bench signal. Another seldom-used implementation method is using a higher-level module to instantiate both the test bench and DUT modules and then tying them together at the higher level.

Test Bench Wrapper

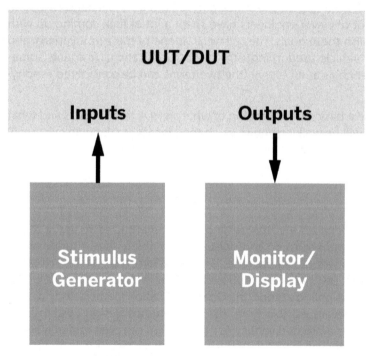

Figure 4-1 *Typical test bench*

Reuse

With the ability to capture hardware designs in a soft form through HDL and archive them in electronic media such as databases came a whole new era for chip design. Inventories of design libraries with hundreds of IP blocks began to spring up within companies, and it was not long before whole corporations began to develop, sell, and support soft IP libraries. Companies like Synopsys (*http://www.synopsys.com/home.aspx*) provide multiple commercial libraries of electronic design elements, from processors to memory, USB and PCI Express (PCIe) I/O controllers, down to simple serial devices like universal asynchronous receiver/transmitters (UARTs) and I2C controllers. Using standard SoC fabric interfaces, like AMBA (Advanced Microcontroller Bus Architecture) and OCP, these IP block products, for the first time, presented a true plug-and-play design methodology. "Reuse" became a buzzword in the design community and the mantra of many hardware design managers. While some engineers cringed at the thought of reusing another designer's IP in their own projects, others welcomed the idea.

Did you ever wonder how a company like Samsung or Apple can pop out an incredibly complicated device like a smartphone every year? They can do it because of reuse. The benefits are obvious. First, the reduction in design turnaround time is tremendous, because major subsystems are on the shelf, so to speak, and ready to go. Also, these elements are tested and typically already proven in other products, so the quality is high. The downside is that the functionality of the elements being reused is somewhat fixed and leaves little room for innovation—which is what most designers want to be doing. In most SoC companies, product differentiation efforts are focused on innovation and developing a so-called special source. (I hope you will be inspired to start developing your own special source IP as a result of reading and using this book.)

But this is not to say that reuse has no place in innovative design. Reuse opens some incredible opportunities to build some pretty elaborate SoCs, given the density that is available today at even the lowest end of the FPGA scale. Fortunately for us, there are open source IP block libraries available free of charge. The one we will be mostly working with in this book is provided by OpenCores (*http://opencores.org*), "The world's largest site/community for development of hardware IP cores as open source."

We'll be talking a lot more about OpenCores in the upcoming chapters.

Running the Test Bench Project

OK, now that we have a bit of background, let's jump right in and learn by example what all this means and how to use it. We are going to use an off-the-shelf IP block from OpenCores for this exercise. First-time users of OpenCores will need to create an account (*http://www.opencores.org*) in order to download any cores from the site.

Step 1: Selection and Download of Core

Once you have created a user account on the OpenCores website, log in and click Projects (Figure 4-2). Then click Communication controller.

Figure 4-2 *OpenCores projects*

In the resulting list, scroll down and click "I2C controller core" under I2C Bus, as seen in Figure 4-3.

HDLC controller	Stats	WDC	
HyperTransport Tunnel	Stats	done	
I2C Bus	Stats		LGPL
⭐ I2C controller core	Stats	done wbc OCCP	BSD
I2C implementation using systemc	Stats		LGPL
I2C Master Core Controller with APB Bridge	Stats		LGPL
I2C Master Slave Core	Stats	done	BSD

Figure 4-3 *Select the I2C core*

Click the Download link under Details. This will download the TAR file *i2c_latest.tar.gz*. If you are using a Windows OS you will need to use a file archive tool like 7-Zip (*http://www. 7-zip.org/*) to unpack this file.

Once the download has completed, move the TAR file to a directory of your choice and use the 7-Zip program to extract the archive. You should now have the directory structure seen in Figure 4-4 on your computer.

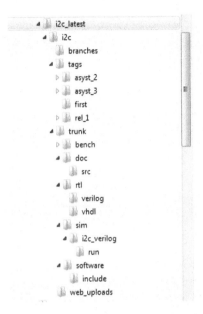

Figure 4-4 *Unpacked I2C core directories*

Step 2: Documentation

This is a good point to review any documentation that has been included with the core. In this example, there is an excellent specification provided in the *doc* directory titled "I2C-Master Core Specification," which includes a great section on the I2C bus's functionality. Open the file *i2c_spec* and read it through.

FYI

This level of documentation is not the norm for open source cores; many of them do not come with any documentation at all.

Did You Know?

For those of you who are not familiar with the I2C bus, it is a low-speed, low-power, two-wire serial bus/protocol that is standard for many peripheral devices used in the electronics and computer industry. The I2C serial bus is a multimaster, multipoint protocol, unlike the very popular UART serial bus protocol, which is point-to-point only. This is one of the big advantages of using I2C over UART. Most of the sensors in tablets and smartphones communicate with the microprocessor core using I2C. If you've used Arduino shields, many of the sensor shields use I2C to communicate with the Arduino. For us FPGA Makers, I2C is a very practical IP block to get familiar with.

*There is a good summary overview of the I2C bus (**http://www.i2c.info/**) online. You can get the full I2C bus standard from NXP (**http://bit.ly/um10204**).*

As you read through the I2C-Master Core Specification provided with this core, pay close attention to Section 2.3, "External connections." The I2C bus uses tristate buffers to implement the serial data line (SDA) and a serial clock line (SCL) for data transfers. This is important to remember when you run the test bench and when you use it in a real FPGA design. You will need to add some logic to your I/O layer for everything to function properly in a real design.

Also notice that this core uses the Wishbone interface as its control/status interface (as described in Section 2.2). Wishbone is an open source hardware computer bus that uses a single simple, high-speed synchronous specification to connect components together in an SoC design. This makes connecting this core to other Wishbone-compliant cores very easy.

FYI

You can find the full Wishbone bus specification online (http://bit.ly/ 1VEaosB).

You will also need to become familiar with Sections 3.1 and 3.2 of the I2C-Master Core Specification, which provide you with the register list and register descriptions. The registers are your interface to the I2C core's operation, and we will be referencing these as we walk through the test bench execution.

What Am I Going to Do with I2C?

At this point, you may be wondering what in the world you are going to do with an I2C core. Suppose you have a project in mind where you want to have your FPGA board connect to an I2C device, like a sensor module. Figure 4-5 shows one possible application of an I2C core in an FPGA design.

In this project the I2C Core is used to communicate with an I2C sensor module using a small state machine to configure and control the core. The information from the sensor is displayed on a 7-segment display. The sensor could be a temperature sensor, for example.

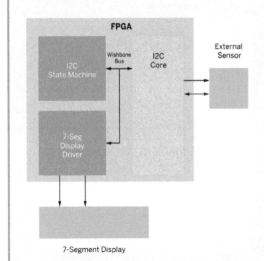

Figure 4-5 *Possible SoC design using I2C core*

Step 3: RTL

Our next step is to get familiar with our design code. I like to pull the RTL files into Xilinx at this point just to see if there is anything quirky in the design code itself before I layer in the test bench code. Also, it gives me a good way to view the design hierarchy using the auto-matic file ordering feature of the tool.

Heads Up!

Just because the code is released to the OpenCores repository doesn't mean it is without errors! You need to check everything, every step of the way. I have learned the hard way not to assume anything.

1. Open a new project in the ISE WebPACK, and then go to Project→Add Source.

2. Navigate to the *i2c/trunk/rtl/verilog* directory of your I2C core, as seen in Figure 4-6, and select all the files (in Windows you can do this by holding down the Ctrl key as you click the files), then click Open.

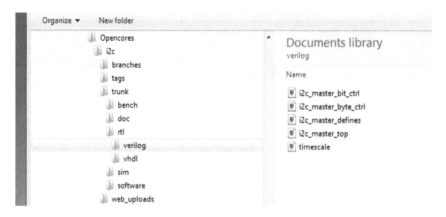

Figure 4-6 *Verilog RTL design files*

3. When all the source files have been added, click OK (Figure 4-7).

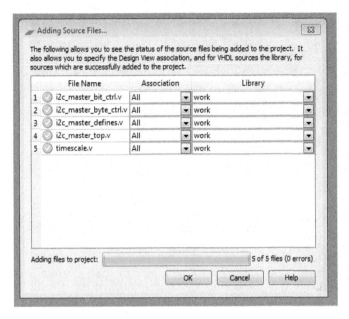

Figure 4-7 *Project source files*

4. Click Implement Top Module (the green triangle on the top toolbar). You should get the "Process Generate Post-Place & Route Static Timing completed successfully" message. This means that you have a clean set (no errors) of RTL design files.

5. At this point I like to take a quick look at the RTL schematic view of the design using the Tools →Schematic Viewer menu option. This gives us a good visual of the I/O pins the test bench will be connecting to (see Figure 4-8). You can also check out the file hierarchy in the Design panel.

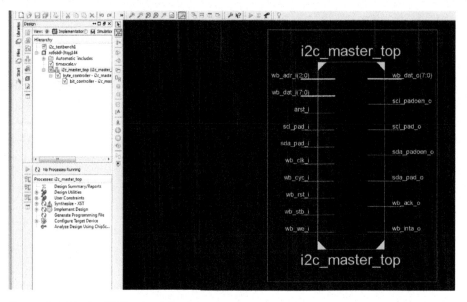

Figure 4-8 *Top-level RTL schematic view*

Step 4: Adding Test Bench Files and Running the Simulation

Now that we have a known good design, it's time to take the big step and bring in our test bench files and run the simulation. We will be using the Xilinx iSim simulator for this exercise. Here's a quick summary:

1. Go to Project→Add Source.

2. Navigate to the *i2c/trunk/bench/verilog* directory of your I2C core and select all the files except the *spi_slave_model* file (I'm not sure why this file is here; it is not used and it also has RTL errors), then click Open (Figure 4-9).

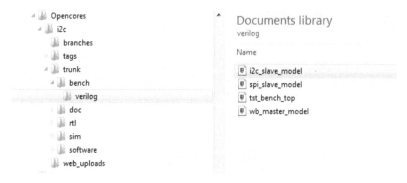

Figure 4-9 *Test bench files*

3. Next, change the Association to Simulation, as seen in Figure 4-10. This is extremely important. If you fail to do this nothing will work and the tool will try to synthesize your test bench, giving you multiple error messages. Then click OK.

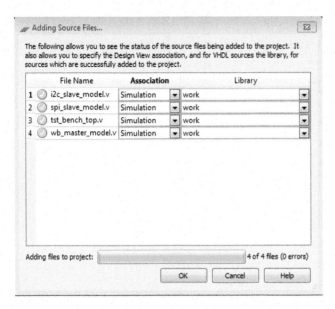

Figure 4-10 *Simulation Association*

4. Switch to Simulation view and you should see the updated hierarchy with the test bench files (Figure 4-11). Notice that *tst_bench_top* is now the top level.

Figure 4-11 *Test bench hierarchy*

5. Next, go to the ISE process windows, expand the iSim simulator, and right-click Simulate Behavioral Model. Click Run.

6. Wait for the iSim tool window to open, then click Simulation and select Run All.

7. Wait for the simulation to finish.

8. Go to the Default.wcfg tab and take a look at the waveform view. Zoom to full view to get a good look at the full simulation run (Figure 4-12).

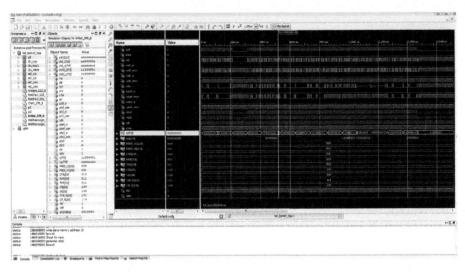

Figure 4-12 *Full test bench simulation run*

9. Switch to the Console view and review the test bench monitor status messages. You should see something like Figure 4-13.

```
Finished circuit initialization process.

INFO: WISHBONE MASTER MODEL INSTANTIATED (tst_bench_top.u0)

status:        5000 done reset
status:      106000 programmed registers
status:      146000 verified registers
status:      176000 core enabled
status:      226000 generate 'start', write cmd 20 (slave address+write)
ISim>
# run all
status:   113416000 tip==0
status:   113466000 write slave memory address 01
status:   214396000 tip==0
status:   214446000 write data a5
status:   409736000 tip==0
status:   409786000 write next data 5a, generate 'stop'
status:   521926000 tip==0
status:   521976000 generate 'start', write cmd 20 (slave address+write)
status:   635136000 tip==0
status:   635186000 write slave address 01
status:   736136000 tip==0
status:   736186000 generate 'repeated start', write cmd 21 (slave address+read)
WARNING: at 745876 ns: Timing violation in /tst_bench_top/i2c_slave/  $setup( scl:742696 ns, sda:745876 ns,4700 ns)


status:   850366000 tip==0
status:   850396000 read + ack
status:   951346000 tip==0
status:   951376000 received a5
status:   951406000 read + ack
status:  1052326000 tip==0
status:  1052356000 received 5a
status:  1052386000 read + ack
status:  1153306000 tip==0
status:  1153336000 received xx from 3rd read address
status:  1153366000 read + nack
status:  1254286000 tip==0
status:  1254316000 received xx from 4th read address
status:  1254366000 generate 'start', write cmd 20 (slave address+write). Check invalid address
WARNING: at 1264036 ns: Timing violation in /tst_bench_top/i2c_slave/  $setup( scl:1260856 ns, sda:1264036 ns,4700 ns)


status:  1368516000 tip==0
status:  1368566000 write slave memory address 10
status:  1469516000 tip==0
status:  1469516000 Check for nack
status:  1469546000 generate 'stop'
status:  1469576000 tip==0


status:  1719576000 Testbench done
```

Figure 4-13 *Test bench monitor status messages*

10. Congratulations—you have run your first test bench. And you didn't have to write one line of HDL code! I told you this would be fun and easy.

Exploring the Test Bench Project

We should have everything we need at this point to begin exploring this IP core using the test bench and documentation provided. We will quickly see how useful a test bench is, and also how easy it is to construct one for our own designs.

Overview

Figure 4-14 presents a simple block diagram to help us identify the components of our I2C Core test bench. I like to have a visual understanding of what is going on as I work through the code. Sometimes the author of the IP will provide a good block diagram, but often this is not the case.

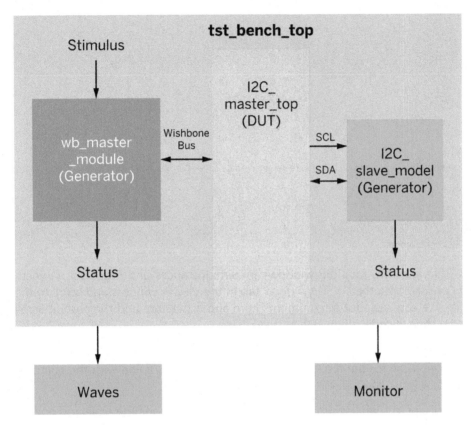

Figure 4-14 *I2C Core test bench block diagram*

Notice that there are two files used as generators for our DUT. *wb_master_module* is our Wishbone bus generator and the *i2c_slave_module* is our I2C slave device generator.

 Did You Know?

The I2C bus protocol is made up of I2C master and slave devices. An I2C master controls the timing of the bus (drives the SCL line), while a slave device only responds to master requests.

Code Walk

OK, let's walk through the *tst_bench_top.v* code to help us understand what is going on. To do this, let's go to our iSim tool (Figure 4-15).

Figure 4-15 *Test bench code*

You'll notice that we have multiple windows in iSim: Instances and Processes, Objects, Viewer, and Console. Click the *tst_bench_top.v* tab in the Viewer window and scroll to the top of the file. We can see a lot of comments here about the core and the author. Scroll down to the first line of code at line #69.

Timescale and delays

At line 69, we see a call for the file *timescale.v*. This file contains a line with the compiler directive *timescale 1ns / 10ps*.

This line is important in a Verilog simulation, because it sets up the time scale and operating precision for the modules. Here, it causes the unit delays to be in nanoseconds (ns) and specifies the precision of the simulator, which, in this case, will round the events down to the nearest 10 picoseconds (ps). This causes a #5 or #1 in a Verilog assignment to be a 5 ns or 1 ns delay, respectively. The rounding of the events will be to .01 ns or 10 ps.

FYI

Delay control is the simplest way to control the timing of a simulation event. It specifies the delay between the time when a statement is encountered and when it is executed. The delay is indicated by the operator # followed by the time to wait.

Instantiating the DUT and generators

As we move down in the test bench top file, we next see the top-level module declaration, much like in our previous projects. Then we see our familiar wires and registers defined.

Before the DUT is instantiated, each of its inputs and outputs must be declared in the test bench. Inputs to the DUT are declared as reg and outputs are declared as wire. Note that the outputs of the DUT are inputs to the test bench. Remember that any input must be a wire. The inputs to the DUT are the stimuli generated in the test bench. Stimuli are usually generated in an initial or an always procedural block in the testbench. We also see a list of parameters—in Verilog, parameters are constants typically used to specify the width of variables and time delays.

We then see the instantiation of the *wb_master_model* at line number 115 and the *i2c_master_top* (our DUT) at line number 136, followed by *i2c_slave_module* at line number 185.

Clocks and resets

Two important elements used in almost all simulations are clocks and resets. All sequential DUTs require a clock signal. To generate a clock signal, many different Verilog constructs can be used. In this case, the always is used:

```
// generate clock
always #5 clk = ~clk;
```

The clock and reset lines are initialized, and then the reset line is toggled by the test bench:

```
// initial values
clk = 0;

// reset system
rstn = 1'b1; // negate reset
#2;
rstn = 1'b0; // assert reset
repeat(1) @(posedge clk);
rstn = 1'b1; // negate reset
```

Display

The Verilog $display is used to print a line to the terminal for viewing or recording by the console. Variables can also be added to the display, and the format for the variables can be set to binary using %b, hex using %h, or decimal using %d. Another common element used is $display is $time, which prints the current simulation time:

```
$display("status: %t done reset", $time);
```

Tasks

Tasks are used to group a set of repetitive or related commands that would normally be contained in an initial or always block. A task can have inputs, outputs, or be bidirectional (in and out) and can contain timing or delay elements.

We can see our first instance of task calls in our code (beginning at line number 235 in our editor view):

```
// program internal registers
      u0.wb_write(1, PRER_LO, 8'hfa); // load prescaler lo-byte
      u0.wb_write(1, PRER_LO, 8'hc8); // load prescaler lo-byte
```

```
u0.wb_write(1, PRER_HI, 8'h00); // load prescaler hi-byte
$display("status: %t programmed registers", $time);

u0.wb_cmp(0, PRER_LO, 8'hc8); // verify prescaler lo-byte
u0.wb_cmp(0, PRER_HI, 8'h00); // verify prescaler hi-byte
$display("status: %t verified registers", $time);
```

To see where the task code originates, click the task in the Instances and Processes window of iSim and then double-click the task *wb_write* (Figure 4-16). ·

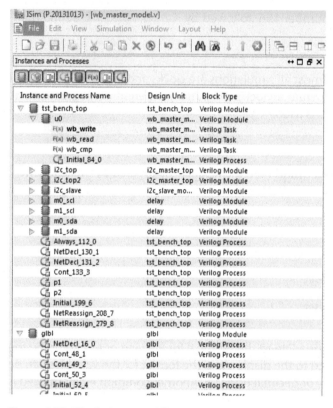

Figure 4-16 *iSim Instances and Processes—select task "wb_write"*

You should now have the code for the task *wb_write* from the *wb_master_model.v* file in your Viewer window. It looks like this:

```
/////////////////////////////////////////////////////////////////////////
//
// Wishbone write cycle
//

task wb_write;
 input    delay;
 integer delay;
```

```
  input [awidth -1:0] a;
  input [dwidth -1:0] d;

  begin

    // wait initial delay
    repeat(delay) @(posedge clk);

    // assert wishbone signal
    #1;
    adr  = a;
    dout = d;
    cyc  = 1'b1;
    stb  = 1'b1;
    we   = 1'b1;
    sel  = {dwidth/8{1'b1}};
    @(posedge clk);

    // wait for acknowledge from slave
    while(~ack) @(posedge clk);

    // negate wishbone signals
    #1;
    cyc  = 1'b0;
    stb  = 1'bx;
    adr  = {awidth{1'bx}};
    dout = {dwidth{1'bx}};
    we   = 1'hx;
    sel  = {dwidth/8{1'bx}};

  end
endtask
```

//

We can see from the code that this task generates a Wishbone write cycle. The syntax of the task is wb_write (delay, address, data). So, in our first task call we had:

u0.wb_write(1, PRER_LO, 8'hfa); // load prescaler lo-byte

This translates to "generate a Wishbone write cycle after a delay of 1 ns, to address 3'b000, with a data field of 8'hFA."

Remember that PRER_LO is a parameter (constant) set above. Our core documentation gives us the register addresses for our DUT, as seen in Figure 4-17.

3.1 Registers list

Name	Address	Width	Access	Description
PRERlo	0x00	8	RW	Clock Prescale register lo-byte
PRERhi	0x01	8	RW	Clock Prescale register hi-byte
CTR	0x02	8	RW	Control register
TXR	0x03	8	W	Transmit register
RXR	0x03	8	R	Receive register
CR	0x04	8	W	Command register
SR	0x04	8	R	Status register

Figure 4-17 *I2C Core registers*

Read Section 3.2.1 of the I2C Core spec (refer back to "Step 2: Documentation") to learn what writing FA hex to the prescaler lo-byte means.

You can do this same exercise for all the remaining task calls to understand what they are doing in the test bench code, reading the register descriptions and explanations of operation provided in the documentation.

Waves

Waveforms are very useful for viewing large amounts of data quickly and efficiently. As you gain experience with FPGA design you will become more and more mindful of the need to look at the waves for your designs. You can also run a simulation and save a set of waveforms and then compare them to waveforms from another simulation run. Graphical analysis is an easy way to see if there is a difference in test results. In Example 4-1, iSim uses the SHM (Simulation History Manager) format for wave data.

Example 4-1 *Wave code*

```
initial
    begin
        `ifdef WAVES
            $shm_open("waves");
            $shm_probe("AS",tst_bench_top,"AS");
            $display("INFO: Signal dump enabled ...\n\n");
        `endif
```

Many formats for wave files exist. The most common is the VCD (value change dump) format.

Stepping

We'll wrap up with a quick look at stepping through the test bench code using break points. Here is a short summary:

1. After your first run through the code in iSim, click Simulation and select Restart.

2. Click the *tst_bench_top* tab in your iSim Viewer window and drop a breakpoint at line 235. A breakpoint is added by clicking the line where you want the breakpoint to be located and then clicking Toggle Breakpoint, as seen in Figure 4-18.

```
197      pullup p2(sda); // pullup sda line
198
199      initial
200        begin
201          `ifdef WAVES
202              $shm_open("waves");
203              $shm_probe("AS",tst_bench_top,"AS");
204              $display("INFO: Signal dump enabled ...\n\n");
205          `endif
206
207  //       force i2c_slave.debug = 1'b1; // enable i2c_slave debug information
208          force i2c_slave.debug = 1'b0; // disable i2c_slave debug information
209
210          $display("\nstatus: %t Testbench started\n\n", $time);
211
212  //       $dumpfile("bench.vcd");
213  //       $dumpvars(1, tst_bench_top);
214  //       $dumpvars(1, tst_bench_top.i2c_slave);
215
216          // initially values
217          clk = 0;
218
219          // reset system
220          rstn = 1'b1; // negate reset
221          #2;
222          rstn = 1'b0; // assert reset
223          repeat(1) @(posedge clk);
224          rstn = 1'b1; // negate reset
225
226          $display("status: %t done reset", $time);
227
228          @(posedge clk);
229
230          //
231          // program core
232          //
233
234          // program internal registers
235          u0.wb_write(1, PRER_LO, 8'hfa); // load prescaler lo-byte
236          u0.wb_write(1, PRER_LO, 8'hc8); // load prescaler lo-byte
237          u0.wb_write(1, PRER_HI, 8'h00); // load prescaler hi-byte
238          $display("status: %t programmed registers", $time);
239
```

Default.wcfg   tst_bench_top.v   wb_master_model.v

Figure 4-18 *Setting a breakpoint in iSim*

3. Now click Run All in the top toolbar. The simulator will stop at the breakpoint you inserted.

4. Now you can single-step through the task code by clicking Step in the top toolbar. Don't forget to take a look at the waveform viewer as you step.

You can now step through the code, adding breakpoints and exploring the operation. Refer to Section 4 of the I2-Master Core Specification, which outlines the operation of the core.

Takeaways

Here are the key takeaways from the exercises in this chapter:

- A test bench is a virtual testing environment used to verify that a design does everything it is supposed to do and doesn't do anything it's not supposed to do.

- A test bench applies stimuli (inputs) to the unit under test (UUT), also referred to as the device under test (DUT). It monitors the outputs, providing status and error reporting in a readable and user-friendly format.

- OpenCores is the world's largest site/community for development of hardware IP cores as open source.

- The I2C bus is a low-speed, low-power, two-wire serial bus/protocol that is an industry standard for many peripheral devices used in the electronics and computer industry.

- Using the documentation provided with the IP core, if any.

- How to make effective use of the Xilinx ISE WebPACK to automatically generate and display the file hierarchy of an IP core.

- Using the Xilinx ISE Simulator (iSim) with a Verilog test bench.

- Adding test bench simulation files to a Xilinx ISE project.

- Simulating an off-the-shelf IP core without writing any code.

- Using a block diagram to aid in our understanding of the test bench.

- Some basic building blocks of a Verilog test bench.

- Effective use of iSim breakpoints and single-stepping to help in our understanding of IP core operation.

It Does Not Compute | 5

What Is a Computer?

Some of us remember hearing the phrase "It does not compute" used by the robot from the hit 1960s television series *Lost in Space*. When it comes to contemplating what a computer really is, I think many of us can honestly say "It does not compute" or even "Danger, Will Robinson!" In this chapter, we will be learning about basic computer architecture using a very cool OpenCores project, and hopefully it *will* compute when we are finished.

The CARDIAC Computer Model

Around the time when our robot friend from *Lost in Space* was hitting the TV air waves, two engineers from Bell Telephone Laboratories, David Hagelbarger and Saul Fingerman, developed the CARDboard Illustrative Aid to Computation, or CARDIAC for short. CARDIAC was a learning aid developed to teach high school students how computers work. The kit consisted of a die-cut cardboard "computer" model and an instruction manual (see Figure 5-1). The cool thing about CARDIAC was its ability to actually function as a slow and basic computer, demonstrating what a computer was in a very simple and interactive way. It had 100 memory locations and only 10 instructions (Figure 5-2). The memory held signed 3-digit numbers (–999 through 999), and instructions could be encoded such that the first digit was the instruction and the second two digits were the address of the memory to operate on (Figure 5-3). The only register in the model was an accumulator.

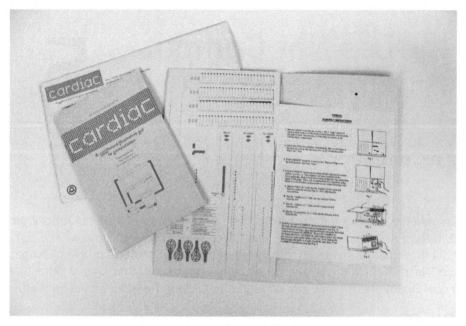

Figure 5-1 *The original CARDIAC cardboard computer kit*

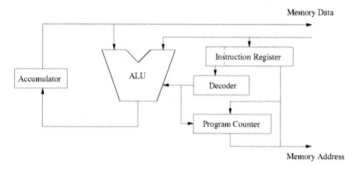

Figure 5-2 *CARDIAC instruction paths*

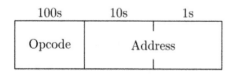

Figure 5-3 *CARDIAC opcode structure*

Brian Stuart of Drexel University has written a great piece on CARDIAC (*https:// www.cs.drexel.edu/~bls96/museum/cardiac.html*), where he explains how the simple instruction set makes for very easy understanding of how complex programs can be built out of simpler sets of operations and data:

> The CARDIAC CPU is a single-accumulator single-address machine. Thus each instruction operates optionally on a single memory location and the accumulator. For example, the ADD instruction reads the data in one memory location, adds it to the current value of the accumulator, and stores the result back into the accumulator. The ALU supports addition, subtraction, and decimal shifting. CARDIAC's CPU architecture is illustrated in Figure 5-4.

Opcode	Mnemonic	Operation
0	INP	Read a card into memory
1	CLA	Clear accumulator and add from memory (load)
2	ADD	Add from memory to accumulator
3	TAC	Test accumulator and jump if negative
4	SFT	Shift accumulator
5	OUT	Write memory location to output card
6	STO	Store accumulator to memory
7	SUB	Subtract memory from accumulator
8	JMP	Jump and save PC
9	HRS	Halt and reset

Figure 5-4 *CARDIAC instructions (source: https://www.cs.drexel.edu/~bls96/museum/cardiac.html)*

You can download the original instruction manual (*http://kylem.net/hardware/cardiac/CARDIAC_manual.pdf*) written by Hagelbarger Fingerman. I think everyone should read this! It is a great way to learn the basics of computer architecture.

At this point you may be wondering, what does this have to do with FPGAs? In the first section of the CARDIAC instruction manual, Hagelbarger and Fingerman write:

> CARDIAC is an acronym for CARDboard Illustrative Aid to Computation. The key word here is "illustrative." It means that CARDIAC illustrates the operation of a computer without actually being a computer. In fact, it is not even a practical aid to computing. On the other hand, it is a very practical aid to understanding computers and computer programming.

> You'll need this kind of understanding to keep up with the Computer Age you are about to enter. These are fast-moving times, and those who make no effort to understand computers may very well get left behind.

These words were almost prophetic—I bet they would never have dreamed back in 1968 that less than 50 years later DIYers like us would be able to actually build a real CARDIAC computer on a $29.95 FPGA board. These *are* fast-moving times!

Getting Started with VTACH

VTACH is an OpenCores FPGA project that is actually a Verilog implementation of the original CARDIAC teaching computer from Bell Labs. The original VTACH implementation runs on a Spartan 3 board from Digilent, and according to its developers "it is pretty faithful to the original."

VTACH is a very easy design to understand. Figure 5-5 shows that the VTACH core is where the CARDIAC arithmetic logic unit (ALU) and 7-segment display driver are located. The input for the core comes from the DIP switches or the push buttons. The main system clock is generated using a DCM, and a block RAM is used for the program memory.

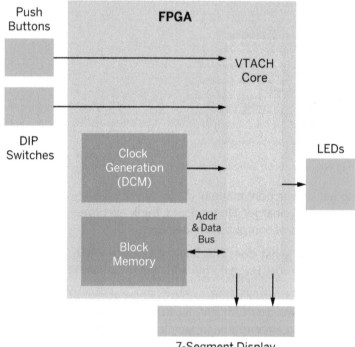

Figure 5-5 *VTACH FPGA design*

You can find the complete VTACH project on the OpenCores website under Project→Processor→VTACH - Bell Labs CARDIAC reimagined in Verilog (*http://open cores.org/project,vtach*).

For this project I will be using the $29.95, Numato Labs, Elbert V2 - Spartan 3A FPGA Development Board (*http://numato.com/elbert-v2-spartan-3a-fpga-development-board.html*).

There are some great web pages that highlight the VTACH design (*http://opencores.org/ project,vtach*) on the VTACH project overview page:

- The Heart of a CPU (*http://ubm.io/1QhvNEk*)
- Expanding VTACH (*http://ubm.io/1QhvN7k*)
- CARDIAC to FPGA (*http://ubm.io/1QhvQ2V*)
- Paper to FPGAz (*http://ubm.io/1QhvRUI*)

You should read them before you move on.

To get started, download the latest version by clicking the download link on the project page (Figure 5-6).

FYI

Remember that you will need to have an OpenCores account to download the project.

Figure 5-6 *Download the VTACH project*

You should now have a ZIP file in your download folder named *vtachspartan.zip*. Move this to a new folder and unzip the file. This project contains a complete Xilinx ISE project image, so we don't need to do much to get it going. How good is that?

Just find the ISE project file in the *vtachspartan* directory and double-click it (see Figure 5-7).

FYI

If you get any warnings about ISE versions and making backups, just click "No Back Up" and continue.

vtachspartan	vtach_tb_ben.prj	PRJ File	1 KB No
_ngo	vtach_tb_stx_beh.prj	PRJ File	1 KB No
_xmsgs	vtach_test	V File	1 KB No
asm	vtach_test_beh.prj	PRJ File	1 KB No
auto_project_xdb	vtach_test_isim_beh	Application	43 KB No
ipcore_dir	vtach_test_isim_beh.wdb	WDB File	237 KB No
iseconfig	vtach_test_isim_beh1.wdb	WDB File	0 KB No
isim	vtachspartan	Xilinx ISE Project	6 KB No
planAhead_run_1	vtachspartan.gise	GISE File	3 KB No
xlnx_auto_0_xdb	webtalk	LOG File	1 KB No
xst	webtalk_impact	XML Document	1 KB No
	webtalk_pn	XML Document	2 KB No

Figure 5-7 *Open VTACH ISE project*

Numato Elbert V2 Setup

Since we are not using the same Digilent board as the original design, we will need to make some minor modifications to the code. You can use the description here as a model to remap the project to whichever board you are using. I chose the Numato Elbert V2, pictured in Figure 5-8, because all the LEDs, push buttons, DIP switches, and 7-segment displays that are needed for the project are already on the board.

Figure 5-8 *Numato Elbert V2*

Setup of this board on Windows was relatively straightforward and the Numato Lab website is clean and well laid out. You just need to select the module of your choice from the Numato Products page (*http://numato.com/products/*) and then select the DOWNLOADS tab (Figure 5-9). You'll find everything you need there to get started.

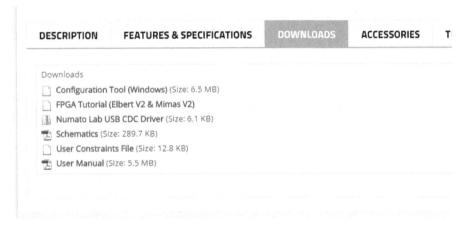

Figure 5-9 *Numato Lab Elbert V2 DOWNLOADS tab*

On Windows platforms, all you need to do is download and extract the CDC driver and config tool to a folder of your choice and then plug the Elbert board into your USB port. In Windows Device Manager, you should see the Elbert listed under "Other devices." Right-click it and select Update Driver Software, then select "Browse my computer for driver software" and point to the folder where you extracted the CDC driver and click OK. Once the driver is installed for the device, you should see the board moved under Ports (COM & LPT). Make a note of which COM port has been assigned to your board, then go to the folder where you extracted the config tool and run it (see Figure 5-10).

Figure 5-10 *Elbert V2 on COM14*

Waxwing

To set up Waxwing (*http://bit.ly/1SguMSF*) on Windows you will need to install a D2XX driver, which can be downloaded from their website (*http://bit.ly/1SguR8V*). You will need to run the CDM v2.08.30 WHQL Certified.exe application (or a newer version), and it will prompt you to install the FTDI CDM drivers. When the driver installation is complete, the module should appear in the Waxwing Flash Config Tool as Waxwing Spartan 6 FPGA Module.

At this point, all you need to do is load the Xilinx FPGA image that is generated from the Xilinx WebPACK ISE tool and you are ready to program your FPGA.

 Setup Test

*You can get complete Xilinx ISE example projects for Elbert V2 from the product page (**http://bit.ly/1SgvGP0**) on the Numato Lab website: example Verilog source files are available under the SAMPLE CODE tab that you can load into a Xilinx project and build. You will need to make sure you specify the correct project settings and have the correct FPGA device selected (see **Chapter 2** for an example flow for building a project).*

Be sure to enable the Create Binary Configuration File option in the Generate Programming File process properties of ISE, as shown in Figure 5-11.

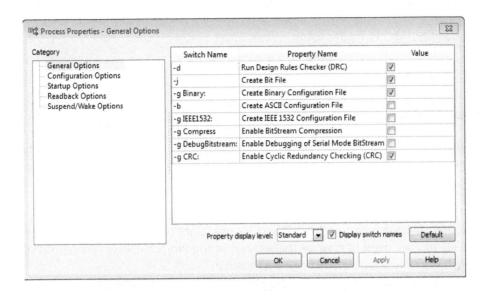

Figure 5-11 *Create Binary Configuration File*

All you need to do next is run the config tool and load your *.bin* file (see Figure 5-12).

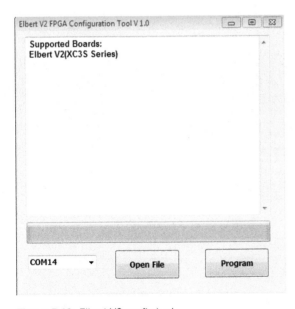

Figure 5-12 *Elbert V2 config tool*

Heads Up!

*Make sure you upload the .**bin** file **not** the bit file!*

Modifications

Depending on your board, the modifications for the VTACH project should be fairly straightforward. I'll review what I did to remap the project to the Elbert V2 in the next few sections.

Step 1: Device Section

Since the original Digilent board and the Elbert V2 use slightly different versions of the Spartan 3 FPGA, you will need to change the settings in the Project→Design Properties of your ISE project. The default project settings are shown in Figure 5-13.

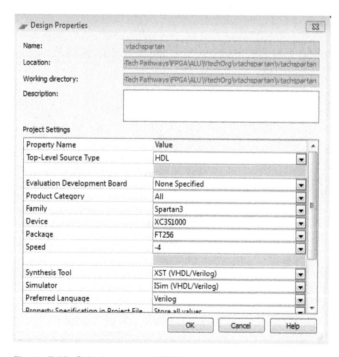

Figure 5-13 *Select new target FPGA*

Checking the Elbert V2 User Guide, we see that it uses the Spartan XC3S50A in the TQG144 package for its FPGA device. We need to add this information to the project settings, as shown in Figure 5-14.

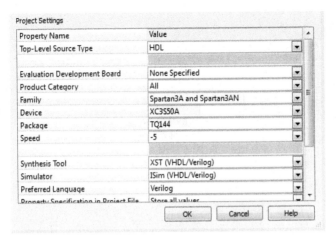

Figure 5-14 *Updated target FPGA for Elbert V2*

Step 2: Pin Assignments

The next thing you will need to get straight are the physical pin mappings for your board and device. As we learned previously, you do this through editing the UCF file in the ISE project. You should see the *vtach.ucf* file in the Design→Implementation→Hierarchy view of the ISE project window (Figure 5-15). Just double-click it to begin editing it.

Heads Up!

You should make a backup of the original UCF file before you edit it.

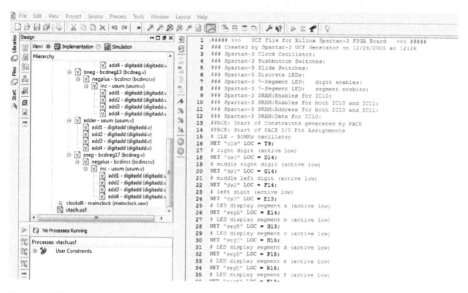

Figure 5-15 *vtach.ucf file*

To make this a little easier, I used the UCF for the Elbert V2 development board that I downloaded from the Numato Labs website as my guide for the pin remapping (see Figure 5-16).

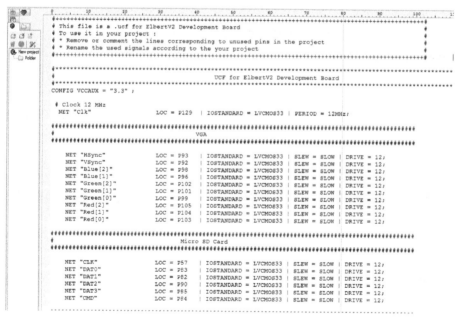

Figure 5-16 *elbertv2.ucf file*

Next, I took the NET names from the original *vtach.ucf* file and plugged them into the matching Elbert V2 pin assignments. Figure 5-17 shows the result. It takes a little bit of cutting and pasting to get all the edits straight, but I think you get the idea.

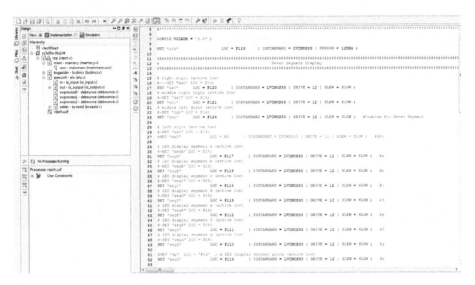

Figure 5-17 *Updated vtach.ucf with Elbert V2 pin assignments*

7-Segment Display

You'll notice that the Elbert V2 board only has three digits of display, not the four that the original Digilent board had. This is not a big deal; I just left the fourth digit enable line (ds3) unconnected.

Step 3: Clocking

Here's where we will need to get into the actual Verilog code and make some modifications. You will notice that in the original *vtach.ucf* file the input clock for the design was operating at 50 MHz.

```
# CLK - 50MHz oscillator
NET "clk" LOC = T9;
```

The Elbert V2 only has a 12 MHz clock available for input into the FPGA:

```
NET "clk" LOC = P129 | IOSTANDARD = LVCMOS33 | PERIOD = 12MHz ;
```

Unfortunately, we cannot use the existing VTACH clocking design with a 12 MHz main clock, the DCM cannot synthesize a 32 MHz clock up from a 12 MHZ clock. I elected to solve this by removing the DCM from the design altogether and just bringing in the 12 MHz clock directly, and then using the existing divide-by-two register to produce the memory divclk. All this means is the design will operate at 12 MHz and 6 MHZ, not the original 32 MHz and 16 MHz. This is really not a big deal for us since we are not all that concerned about performance anyway.

To change the clocking design, you will need to open the top (*vtach.v*) file and edit a couple of lines of code. First, you need to change the `assign` statement for clk2 from `clkls` to clk:

```
assign clk2=clk; // Original vtach was 32MHz, Elbert V2 is 12MHz
```

Then you will need to comment out the instantiation of the `mainclock` DLL:

```
// Instantiate the DLL clock module
//Not used for Elbert V2 Development Board
/*mainclock clockdll (
    .CLK_IN1(clk),
    .CLK_OUT1(clkls),
   .CLK_OUT2(clkdiv),
   .RESET(1'b0)    );
*/
```

The last thing you need to do is right-click the `mainclock` module in the Implementation view in the Design panel of your ISE project and remove it.

Heads Up!

I had some issues getting VTACH to run on an LX9-based FPGA. This may have to do with the skew between the 12 MHz and 6 MHz clocks created by the clock divider FF. Also, you need to be sure that a clock buffer is used on the 6 MHz clock.

Step 4: I/O Polarity

Now that we have the clocking design ironed out, we need to make some modifications to some of the other I/O polarity coding to match the functionality of the Elbert V2 board. We will begin by looking at the reset signal and the push button that controls it. Again, notice that the original UCF file has the external reset signal connected to push button 3 and it is active high:

```
# BTN3 (active high)
NET "extreset" LOC = L14;
```

The Elbert board's push buttons will drive low when pushed, not high (see Figure 5-18). This will cause problems if we just connect the existing logic to the push buttons, as we did in our UCF file modification.

2

2

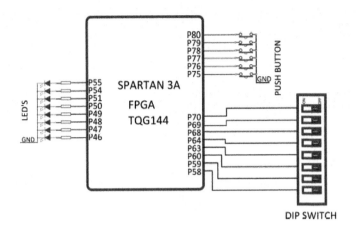

Figure 5-18 *Elbert V2 push button hardware*

We can solve this problem by simply adding an inverter to the original input signal and creating a new active low reset pin at the top (*vtach.v*) level with a pull-up, as illustrated in Figure 5-19.

FYI

*Inputs have optional pull-ups that users can make use of (**Figure 5-19**).*

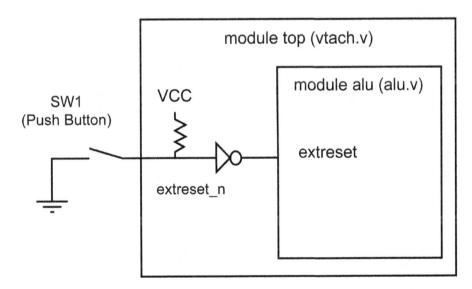

Figure 5-19 *Elbert V2 push button inverter circuit*

I did this for the other push buttons as well, as illustrated in Examples 5-1, 5-2, and 5-3.

Example 5-1 *Adding active low push button pin to top module (vtach.v)*

```
// This is the main CPU module
// Inputs are the main clock, external reset, push buttons, and switches
// Outputs are the 7-segment display (segX and dsN) and the 8 discrete LEDs
module top(input clk,input extreset_n, output segA, output segB, output segC,
  output segD, output segE, output segF, output segG, output segH,  output ds0,
  output ds1, output ds2, output ds3, output [7:0] led,
  input pb0_n, input pb1_n, input pb2_n, input [7:0] sw);
```

Example 5-2 *Adding inverters top module (vtach.v)*

```
assign extreset=~extreset_n;
 assign pb0=~pb0_n;
 assign pb1=~pb1_n;
 assign pb2=~pb2_n;
```

Example 5-3 *Adding pull-up function to pin (vtach.ucf)*

```
###########################
Push buttons switches
###########################

# BTN3 (active high)
#SW1
#-NET "extreset" LOC = L14;
NET "extreset_n"  LOC = P80    | IOSTANDARD = LVCMOS33 | DRIVE = 12 | SLEW = SLOW |
PULLUP;

# BTN2 (active high)
#SW2
#-NET "pb2" LOC = L13;
NET "pb2_n" LOC = P79      | IOSTANDARD = LVCMOS33 | DRIVE = 12 | SLEW = SLOW |
PULLUP;

# BTN1 (active high)
#SW3
#-NET "pb1" LOC = M14;
NET "pb1_n" LOC = P78      | IOSTANDARD = LVCMOS33 | DRIVE = 12 | SLEW = SLOW |
PULLUP;

# BTN0 (active high)
#SW4
#-NET "pb0" LOC = M13;
```

```
NET "pb0_n"  LOC = P77        | IOSTANDARD = LVCMOS33 | DRIVE = 12 | SLEW = SLOW |
PULLUP;
```

Step 5: Memory Block Update

The last thing you will need to do is rerun the IP wizard and regenerate the memory block. Follow these steps:

1. Remove the existing `mainmem` module by right-clicking the module in the Design panel and then selecting Remove, as seen in Figure 5-20.

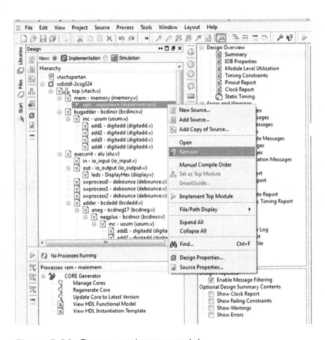

Figure 5-20 *Remove mainmem module*

2. Regenerate the `mainmem` module by selecting "mem - memory (memory.v)" in the Design panel and then, from the toolbar, selecting Project→New Source.

   Give it the same filename as was originally used (*mainmem*), click IP (CORE Generator and Architecture Wizard), and then click Next. Click Yes in reply to the questions that follow; then select Memories & Storage Elements→RAMs & ROMs→Block Memory Generator (see Figure 5-21) and click Next, followed by Finish.

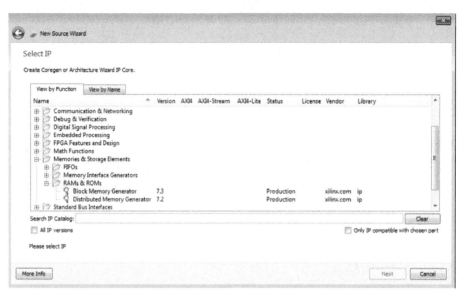

Figure 5-21 *Block Memory IP Generator*

3. Click Next for the first two pages of the Block Memory Generator, then on page 3 (Port A Options), change the Write Width to 13, the Write Depth to 100, and select Read First, as shown in Figure 5-22. Then click Next.

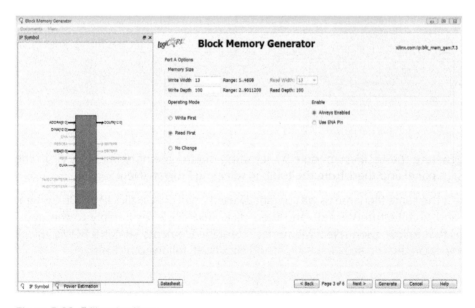

Figure 5-22 *Edit port options*

4. On page 4 (Memory Initialization), check Load Init File and browse to the *mainme-mory.coe* file in your *vtachspartan* directory, as seen in Figure 5-23.

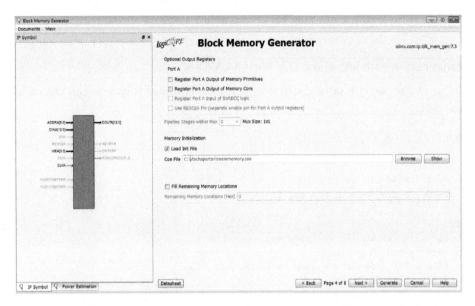

Figure 5-23 *Add memory initialization file*

5. The *mainmemory.coe* file (reproduced in Example 5-4) contains the actual program code that will run on your VTACH computer.

Example 5-4 *COE file for testing of VTACH computer*

```
;Generated by soloasm
MEMORY_INITIALIZATION_RADIX=16;
MEMORY_INITIALIZATION_VECTOR=
0480,
0670,
0570,
0490,
0308,
0409,
0314,
0802,
0170,
0220,
0670,
0490,
0311,
0802,
0170,
0720,
```

```
0670,
0409,
0317,
0802,
0001;
```

6. The numbers in each line of the COE file is a CARDIAC.

Example 5-5 shows the same program with some comments so you can see what is going on.

Example 5-5 *The COE file with comments*

```
0- 0480 - Read Dip Sw to ACCU
1- 0670 - Store Accu to Mem Loc 70

top:
2- 0570 - Write Loc 70 to 7-Segment
3- 0490 - Read PB
4- 0308 - TAC Test Accu & Jump to 8 (exe) is neg
5- 0409 - Read PB (-1 or 1)
6- 0314 - TAC Test Accu & Jump to 14 is neg
7- 0802 - Jump to 2 (TOP) and save PC

exe:   ; add 1 to count

8- 0170 - Clear accumulator and add from memory (load)
9- 0220 - Add from memory to accumulator
10- 0670 - Store Accu to Mem Loc 70

wait:   ; wait for button up
11- 0490 - Read PB
12- 0311 - TAC Test Accu & Jump to 11 (wait) is neg
13- 0802 - Jump to 2 (top) and save PC

exe2: ; decrease count
14- 0170 - Clear accumulator and add from memory (load)
15- 0720 - Subtract memory from accumulator
16- 0670 - Store Accu to Mem Loc 70

wait2: ; wait for button up
17- 0409 - Read PB
18- 0317 - TAC Test Accu & Jump to 17 (wait2) is neg
19- 0802 - Jump to 2 (top) and save PC

one:
20- 0001 - Read a card into memory
```

7. Finish up by clicking Next and then Generate on page 6. At this point you are ready to implement the design and simulate it.

Design, Build, and Simulation

To build your design with these modifications, you just need to click the green triangle in the toolbar (Implement Top Module) as we have done before. If you encounter any errors, you will need to check your edits for syntax and make sure you have the right pin numbers assigned in the UCF.

Once you have a clean build (you will have some warnings, but you can ignore these), you are ready to simulate.

Simulation

Before you can run the simulator you will need to make some changes to the test bench to accommodate the polarity changes we made for the push buttons. Switch to the Simulation view in the Design panel and you'll notice that there are two test bench files (*bcadd_tb* and *vtach_test*). We are only interested in *vtach_test* for now. The other is a simulation of the BCD (binary-coded decimal) math conversion algorithm, which you can run later if you are interested in seeing how that works.

Double-click *vtach_test* and change all the push button logic, reflecting the active low designation that you made in the *vtach.v* design file. The file should look like Figure 5-24 when you are done.

```
24
25   module vtach_test;
26
27       // Inputs
28       reg clk;
29       reg extreset_n;
30       reg pb0_n;
31       reg pb1_n;
32       reg pb2_n;
33
34       // Instantiate the Unit Under Test (UUT)
35       top uut (
36           .clk(clk),
37           .extreset_n(extreset_n),
38           .pb0_n(pb0_n),
39           .pb1_n(pb1_n),
40           .pb2_n(pb2_n)
41
42       );
43
44       always #1 clk=~clk;
45
46       initial begin
47           // Initialize Inputs
48
49           clk = 0;
50           extreset_n = 0;
51           pb0_n = 0;
52           pb1_n = 0;
53           pb2_n = 0;
54
55           // Wait 100 ns for global reset to finish
56           #100 extreset_n=1;
57           pb0_n = 1;
58           pb1_n = 1;
59           pb2_n = 1;
60           // Add stimulus here
61
62       end
63
64   endmodule
65
```

Figure 5-24 *VTACH test bench edits*

Now save your changes and click Simulate Behavioral Model. We won't get into constructing a more elaborate test bench here; we will just force a few signals to get the simulation working. You can experiment with developing the test bench code further if you like as one of your own projects.

To get a simulation working correctly, follow these steps:

1. Once your iSim window is open, click the Restart button in the top toolbar.

2. In the Instances and Processes window on the left, right-click *uut* and select "Add to Wave Window" (see Figure 5-25).

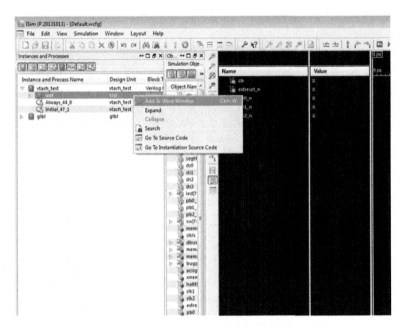

Figure 5-25 *Add uut waves*

3. Now go to the bottom of the wave window, right-click the signal clkdiv, and, as shown in Figure 5-26, force the clock to have a leading edge value =1 of trailing edge value of 0, and a period of 4000. Set Starting at Time Offset to 1010 and click OK. This is required because there is a slight problem with the way the clock divider design is implemented and the way iSim sees it without a delay through the register. This is a shortcut hack, but sometimes it's okay to do this because we know that the design works in the real logic. It didn't seem worth the effort to try to fix it for simulation.

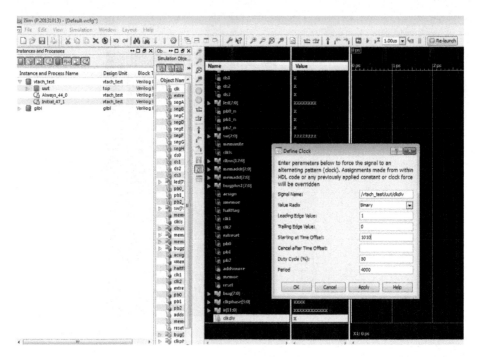

Figure 5-26 *Force clkdiv*

4. Next, find the sw[7:0] bus. This is the input from the dip switches. Right-click it and select Force Constant. Set Force to Value to 07 Hex.

5. Click Run in the top toolbar and let the simulator run for a few seconds, then click Break.

6. Click the *Default.wcfg* tab, zoom to full view, and you should see the 7-segment display output changing state and putting out the right codes for the 07 Hex read from the dip switches (Figure 5-27).

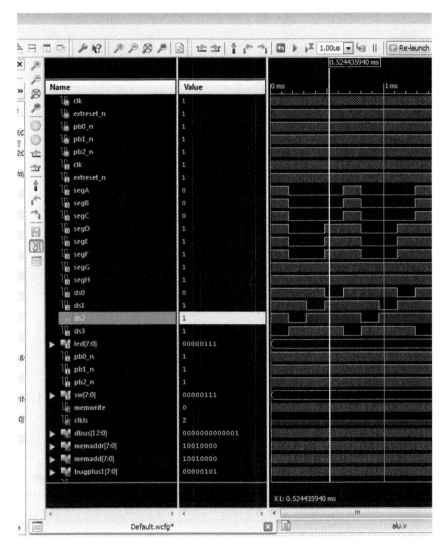

Figure 5-27 *7-segment display*

7. Change the dbus, memaddr radix to hex and zoom in until you can see your program codes on the dbus.

8. Select the memory tab, double-click the block memory link, and you will see the contents of your memory, which includes your program and data (location 70)—see Figure 5-28.

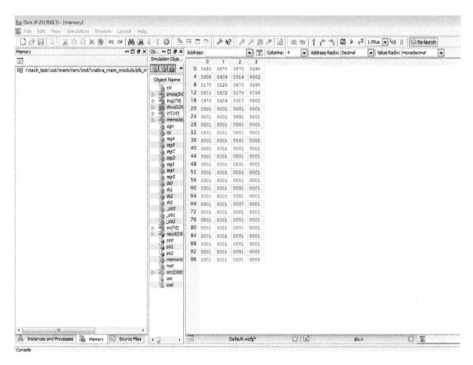

Figure 5-28 *Memory display*

Now that it looks like everything is working, we just need to create the bit file and upload it to our board.

Building and Running

Go back to the Implementation view of your ISE project in the Design panel and select the top module of your hierarchy. Right-click Generate Programming File and select Process Properties. For the Elbert V2, we need to make sure that the -g Binary option in the General Options category is selected. Once that's done, click OK.

Now you can right-click Generate Programming File and select Run.

When the process is complete, upload the *.bin* file to the Elbert V2 board, following the procedure in the setup chapter.

Once that's completed, you should see the BCD number that you have set on your dip switches displayed on your 7-segment display (see Figure 5-29).

Figure 5-29 *VTACH running on Elbert V2*

Now you can go through and test the program's functionality. Change the dip switches and push SW1 (reset), and you should get a new number on the 7-segment display. Push SW2, and the number decrements; pushing SW4 increments the number.

Congratulations! It computes!

Programing and Assembler

Now you can really start to experiment by writing your own CARDIAC programs and running them on your VTACH computer. To do this, you will need to create new *.coe* files, load them into the memory using the IP wizard (refer back to "Step 5: Memory Block Update"), and then resynthesize by clicking Implement Top Level in ISE. Then you will need to generate a new bit file.

Al Williams has written a universal cross assembler called *axasm* that you can use to test the VTACH. It also generates a *.coe* file for you. You can download the assembler from Dr. Dobb's website (*http://ubm.io/1JWr90P*).

On Windows, you will need to do the following:

1. Install Cygwin with *gcc* and *awk* packages.

2. Create a directory with your files.

3. Add Cygwin to your path.

4. Open a DOS window.

5. Type **sh** to get to a Cygwin shell.

6. Run your *axasm* with *./axasm*.

FYI

*On Linux, this is more straightforward; just run **axasm** from a terminal window.*

Example 5-6 shows the program used to test the VTACH, written in assembler.

Example 5-6 *mainmemory test program assembly code*

```
;;      Test program to debug the BCD math
ORG 0
SFT 8,0             ; read switches to AC
STO 70
top:
OUT 70
SFT 9,0             ; read push button (-1 or 1)
TAC exe             ; button down?
SFT 0,9             ; read other push button
TAC exe2
JMP top
exe:                ; add one to the count
LOD 70
ADD one
STO 70
wait: SFT 9,0       ; wait for button up
TAC wait
JMP top

exe2: LOD 70        ; subtract one from the count
SUB one
STO 70
wait2: SFT 0,9      ; wait for button up
TAC wait2
JMP top

one: DATA 1
END"
```

You really don't need to mess with the assembler program to get started. I just copied and renamed the *mainmemory.coe* file, then started writing my own instruction list in the *.coe* format.

Have fun!

Takeaways

Here are the key takeaways from this chapter's exercises:

- The CARDboard Illustrative Aid to Computation, or CARDIAC for short, was a learning aid developed by David Hagelbarger and Saul Fingerman in 1968 to teach high school students how computers work. The kit consisted of a die-cut cardboard "computer" model and instruction manual.

- VTACH is an OpenCores FPGA project that is actually a Verilog implementation of the original CARDIAC teaching computer from Bell Labs.

- Setup of the Numato Elbert V2 board on Windows is relatively straightforward, and the Numato Lab website provides everything you need to get started.

- You can really start to experiment by writing your own CARDIAC programs and running them on your FPGA VTACH computer.

It's a Small World!

§

Easy System-On-Chip (SoC) Designing

To many of you it may seem like a daunting task to tackle a real SoC design in an FPGA. We have seen how HDL can be used to make the design task more manageable, but there's still a learning curve. Schematic entry provides us with a graphical design entry method, but it typically doesn't scale well for large, complex designs. Lucky for us, there is a hybrid approach that combines the best of both worlds and simplifies the SoC design task. The method involves using a hierarchical approach that begins with schematic entry for the top level, which connects our IP blocks together, and then uses HDL at the next level down to describe the function of each IP block. You can think of this as sort of a functional block diagram approach. While this approach is not used much in industry, it is a great approach for us DIYers, because the interconnection of IP blocks can be done graphically.

In this chapter, we will be exploring an SoC design of a VGA video and YM2149 audio player project made easy with DesignLab and the Papilio DUO.

System on Chip

In recent years, with the advent of tablets and smartphones, you may have seen and heard a lot about SoCs—so what is an SoC, and what is all the hoopla about? Simply put, an SoC is a semiconductor microchip ("chip" for short) that contains multiple electronic components integrated together on a single silicon die. This single chip may contain digital, analog, mixed-signal, and even RF (radio frequency) functions that collectively comprise a complete system. We will be focusing on digital logic SoC designs. From Webster's Dictionary, we get the definition of "system" as "a set of interacting or interdependent components forming an integrated whole." The key here is the phrase "integrated whole." The SoC is used and viewed as one integrated whole electronic device, even though there are many subfunctions represented within the system. A car can be used as a good analogy—it's a complete system made up of many subsystems integrated into one platform. For example, you have the fuel, braking, electrical, engine, and drive systems all integrated and functioning together to make one complete car.

With the massive volume of resources (transistors) that have become available on a single silicon die over the last few decades, the idea of integrating more functions on a single chip was a natural progression. In the past, multiple printed wiring boards (PWBs) were used to connect computer subsystems—for example, sound cards, memory modules, graphics processing units, networking controllers, etc. Today, in a typical tablet or smartphone, most of these subsystems are all integrated onto a single SoC chip (Figure 6-1).

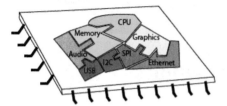

Figure 6-1 *Integrated smartphone systems on a single chip*

The SoC industry is a direct result of Moore's Law and the integrated circuit (IC) manufacturing capability that follows it. In 1965, Gordon Moore, cofounder of Intel Corporation, observed that the number of transistors per square inch on integrated circuits had doubled every couple of years since the IC was invented. Moore predicted that this trend would continue for the foreseeable future.

Moore's Law

The number of transistors that can be packed onto a microchip doubles every two to three years.

A Short History Lesson

This information may be handy next time you play Trivial Pursuit. In 1958, Jack Kilby of Texas Instruments developed the first IC. It contained only five components, including transistors, resistors, and capacitors. By 1997, the Intel Pentium II, with a clock speed of 233 MHz, contained over 7.5 million transistors. Think of that! In fewer than 20 years, the integrated circuit evolved from a couple of transistors on a crude silicon substrate test bed to 7.5 million transistors on a tiny silicon die. By 2014, the Intel i5 quad-core processor, with a clock speed of 2.9 GHz, contained over 1.4 billion—with a "B"!—transistors. That's the exponential increase in density Gordon Moore predicted five decades ago.

Just like the cell is the basic building block for the systems in the human body, the transistor is the basic building block for computer systems. Think of the transistor as a switch that is either "on" or "off," "one" or "zero." Hey, that's binary...so that's where all those 1s and 0s come from! Transistors are combined to form digital logic circuits, which are combined to build digital systems. Resistor–transistor logic is a form of logic structure that IC designers use to create logic functions like TTL (*http://bit.ly/1SgBsjJ*) and CMOS (*http://bit.ly/1VE5wDA*). Think of a modern computer chip, with over a billion switches turning on and off, over a billion times a second—it makes your head spin! If you ask, how can you even begin to connect a billion of anything? Well, hang in there; we're getting to that.

SoC Architecture

To help with your understanding of the FPGA designs we will be exploring in this and the following chapters, you'll need a little bit of background information on the architecture of SoCs. Let's look at a practical example. A block diagram of a typical SoC that could be implemented in a smartphone might look something like Figure 6-2.

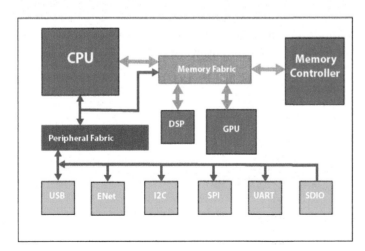

Figure 6-2 *Smartphone block diagram*

Notice that the architecture of the SoC is built around two fabrics: the memory fabric and the peripheral fabric. In the SoC world, a `fabric` is nothing more than an interconnect structure or matrix. These fabrics can be in the form of point-to-point buses, crossbar switch matrices, or even sophisticated packet-switched networks. Many times these interconnect fabrics are generated by computer-aided design (CAD) tools rather than being coded by hand. There are a couple of fabric standards that most commercial SoC vendors use. The Advanced Microcontroller Bus Architecture (AMBA) is used mostly with ARM-based SoCs and is the fabric that you find in many popular ARM-based smartphones and tablets today. The other major player in fabric technology for SoCs was created by the Open Core Protocol (OCP-IP) organization, which started work on what became the OCP specification. These two standards were both meant to solve pretty much the same problem. The idea is that through these fabric standards, designers are able to connect complicated SoC subsystems together on a single chip.

There are other players in the SoC fabric game, of course, and some SoC vendors even develop their own fabrics because of special functionality requirements, or just to be different. The fabric that will be most interesting to us as noncommercial SoC experimenters and designers is the Wishbone Bus Interface Standard. Wishbone is an open source hardware computer bus that uses a single simple, high-speed synchronous specification to connect components together in an SoC design. Since it is open source, hobbyists like us can use the Wishbone bus free of charge. Simple and free; what's better than that? More on Wishbone is coming up later. You can find the full specification on the OpenCores website (*http://bit.ly/1VEaosB*).

Like SoCs, chassis-based platforms are used to expand the capability of computer systems. Here, a backplane of parallel connections forms a "bus" (often referred to as a "rack") that many different cards can be plugged into, as seen in Figure 6-3. A system designer can choose cards from a catalog (library) and simply plug them into the rack to get a customized system with a processor, memory, and interfaces appropriately selected for the given application. Many high-end network servers are configured like this.

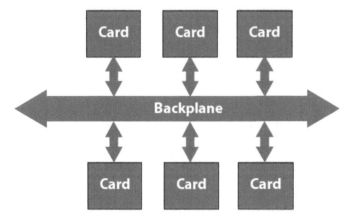

Figure 6-3 *Rack-based computer system*

Similarly, today a designer of an SoC can select design blocks, place them onto a chip, and connect them with a standard on-chip fabric like AMBA or OCP. The backplane might not be apparent as a set of parallel wires on the chip, but logically the solution is the same (see Figure 6-4). SoC designers typically call these blocks *intellectual property blocks*, better known as *IP blocks*. The name reflects the nature of the element, in that it is a soft design element, usually coded in a hardware description language, like VHDL or Verilog. In its library or catalog state, the IP block is not implemented into hardware yet, so it's really just intellectual property, like a source code file for a software program in a sense. You can think of an IP block as a software file that turns into hardware through the magic of synthesis.

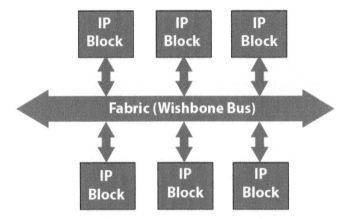

Figure 6-4 *SoC IP blocks and fabric*

 FYI

I'll be using the Papilio DUO exclusively for these projects, but with a little work you can remap these designs to other FPGA boards. You may need to create some additional breadboard circuits to get the I/O you need, though, which the Papilio DUO Computing Shield provides (this topic is beyond the scope of this book).

We have seen that there are many IP cores available to us on sites like OpenCores (*http://opencores.org*), but how do we go about hooking them together easily? The folks at Gadget Factory (*http://www.gadgetfactory.net*) have taken the hybrid design entry approach to a whole new level within their DesignLab tool, and I must admit, it is really cool and I love it. I've used the hybrid approach in my own FPGA design career with great success, so I'm excited to see it being exploited in DesignLab. I know that some of you HDL purists are already turning off, but stay with me, I may even convert you.

DesignLab

The Papilio DesignLab IDE (*http://bit.ly/1VEaAb7*) is an attempt by the folks at Gadget Factory to make FPGA design easier for their Papilio (*http://papilio.cc/*) FPGA development platforms:

> Our dream is to take the hardcore out of FPGA (Field Programmable Gate Array) and make it an amazing tool that anyone can use for creative technology projects.

Some of you may still be struggling to get to grips with HDL, but it's really not that bad once you get used to some of the particulars.

DesignLab provides a drag-and-drop interface for designing FPGAs using the Arduino IDE, adding circuit constructs into it. The tool is only good for use with the Papilio boards, so it is very limited in that regard.

If you are using a Papilio board, you can download the DesignLab IDE from the Downloads section of the Gadget Factory website (*http://bit.ly/1VEaCzW*).

Installation

The installation of DesignLab is easy: you just need to pick either a Windows or a Linux download. The Windows download provides you with an easy *setup.exe* file, while the Linux 32 and 64 downloads provide a tarball that contains a setup script.

 Heads Up!

To use DesignLab you will also need to install the free Xilinx ISE WebPACK software, as described in **Chapter 2***.*

Here's a short summary of the install process:

1. To download the latest version of the DesignLab IDE, go to the Gadget Factory download page, click Agree & Download. If you're using Linux, be sure to select the correct version (linux32 or linux64), depending on your operating system.

2. Extract the *TAR* file to a location of your choice.

3. Click *setup.exe* or run the install script for Linux.

Heads Up!

*The Linux install script was created and tested on Ubuntu Linux. It may work on other Debian derivatives, but for other flavors of Linux you will need to install the required packages by hand. Refer to the DesignLab installation guide (**http://bit.ly/1R8vOwC**) for details.*

Papilio DUO Setup

Heads Up!

Please note that due to the rapidly changing world of FPGA development boards and development technology, the setup procedures described here are subject to change. The following procedures were accurate at the time of writing but may have changed since.

To help you visualize the default configuration of the DUO platform with respect to the USB connectors (PWRSEL jumper and user switch SW1), I came up with the block diagram in Figure 6-5.

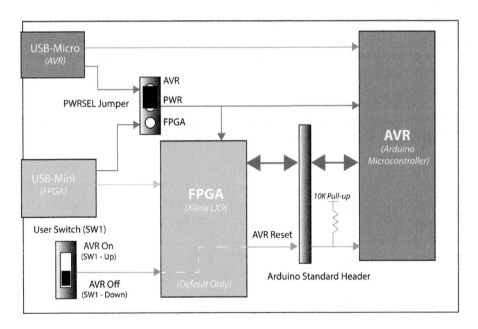

Figure 6-5 *Simplified block diagram of default Papilio DUO platform*

The DUO hardware design is actually pretty straightforward. Power for the module comes from either the USB-Micro connector (top-left), which also provides a communication link to the AVR (Arduino microcontroller), or the USB-Mini connector (bottom-left), which also provides a communication path to the FPGA. The PWRSEL jumper (3-pin header) provides the power source selection. My board was shipped with the jumper connecting the middle (power) and top (AVR) pins. In this configuration, power is being sourced from the USB-Micro connector. If you move the jumper block to the middle and bottom pins, then power will be sourced from the USB-Mini connector (FPGA). The reset signal to the AVR is connected to the user switch (SW1) by an initial (factory) default FPGA configuration. If SW1 is in the "up" position, then the AVR is "on" (reset is inactive). When SW1 is in the "down" position, the AVR is "off" (reset is active). You must understand the routing of FPGA pin 139 (AVR reset) in your particular FPGA design when you load or create a new bit file (overwriting the factory default). If pin 139 is set to tristate, then AVR reset is pulled high and the AVR is on. Clear as mud, right? This will become more clear as we move along.

Okay, now we're ready to load our first FPGA sample design using the DesignLab IDE to test our setup before we begin our project.

Step 1: Power Up

At this point, you need to be sure you have installed the Xilinix ISE WebPACK on your computer. You can download it from the Xilinx website (*http://bit.ly/1JWwCVt*). If you are installing on a 64-bit Windows 8 computer, be sure to follow the workaround instructions (*http://bit.ly/1NTYlIS*) for the license bug.

 FYI

*See **Chapter 2** for more details on the ISE WebPACK.*

I first moved the PWRSEL jumper to the bottom position, selecting the source of power to be the USB-Mini connector. I then plugged in the USB cable from my laptop and the board powered up, which was indicated by the red LED coming on.

Step 2: Select COM Port

Next, you will need to select the port that is connected to your Papilio DUO FPGA from the Tools menu of DesignLab. In my case, it was COM6. You should see it clearly labeled in the Port submenu (Figure 6-6). It should look something like "COM6 (Papilio DUO FPGA)."

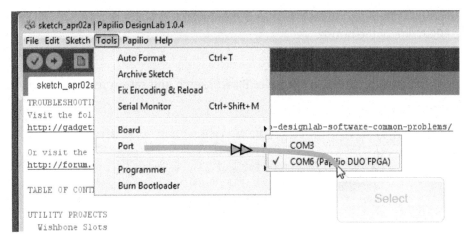

Figure 6-6 *Select your DUO FPGA port*

You also need to select your board from the Tools menu. In my case, I selected Papilio DUO FPGA 2MB ZPUino from the Board submenu because this is the physical board type that I used (Figure 6-7).

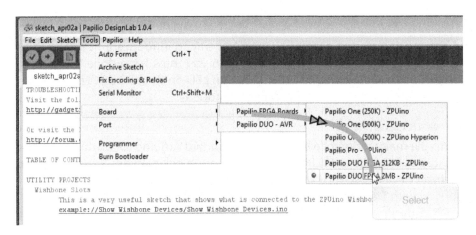

Figure 6-7 *Select your DUO FPGA board*

Heads Up!

It is important that you have the right Tools config selected (port and board type) for the side you're working with (FPGA or AVR)!

I had some issues remembering to keep the port and board type straight in the Tools config setup tabs. This becomes really important when you have both Papilio DUO USB ports connected to your PC. More on this in a moment; for now, to make life simpler, we will just focus on the FPGA side of the house.

Step 3: Create Project

The next step is to open the Arduino example "Blink" sketch and save it as a new project (Figure 6-8).

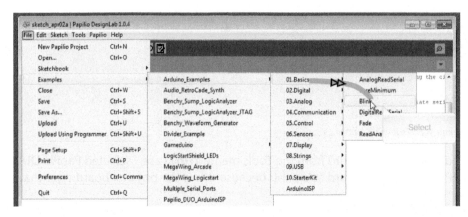

Figure 6-8 *Open the Arduino Blink sketch*

Using "Blink" as a getting started example is a typical way to introduce a new user to an Arduino platform. If you're familiar with Arduino, then you must have done this at least once before. If not, it is a simple program that blinks an LED on the board. It's an easy way to make sure your toolchain is set up and configured properly.

Step 4: Associate Circuit

Now we get into the major difference between the standard Arduino IDE/Arduino module and the DesignLab IDE/Papilio DUO. Remember from our block diagram (Figure 6-5) that there are two main devices on this Papilio module: the FPGA and the AVR (Arduino). The key when working with DesignLab IDE and the DUO module is that you need to always associate an FPGA design with the sketch you are working with. A sketch is simply an Arduino C program source code file. Since we're working with an FPGA, we have to also plug an FPGA design source file into the build environment as well. The way this is accomplished is through the use of a #define preprocessor directive from the C programming language within the sketch (source code) file. In the C programming language, the #define directive allows the definition of macros within your source code. The syntax for this, in the context of DesignLab, is:

```
#define circuit <name>
```

Notice that "circuit" is the keyword of the directive and is followed by the name of the FPGA design bit file. The bit file we are going to use for our first quickstart case is the *ZPUino_Vanilla* FPGA sample design.

This brings us to the next point of clarification. The ZPUino is basically what we call in the industry a "soft design" of the Arduino microprocessor core. In other words, the digital logic of the Arduino core is captured, then synthesized, and becomes a usable library block for FPGA designers. The ZPUino behaves as an Arduino microcontroller, but from inside the FPGA. I covered these concepts earlier in this chapter; you can also find a lot of information about the ZPUino in the User Guide on the Papilio site (*http://bit.ly/1VEaEb5*).

For now, just think of this as another Arduino that you can load and execute sketches on from inside the FPGA. If you take a look at the block diagram in Figure 6-9, you'll notice that the ZPUino and AVR are both connected to the Arduino standard shield header. This is why you need to either turn the AVR off by putting SW1 down or pay close attention to what pins each is using in your FPGA design and AVR code to avoid contention on the Arduino shield connector. We will just be turning AVR off (SW1 down) for this exercise.

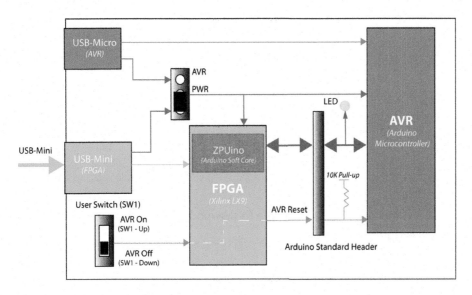

Figure 6-9 *Papilio DUO FPGA with ZPUino added, ready for Blink test*

We now need to add the following line of code to our Blink sketch, as shown in Figure 6-10, and save it as a new file:

```
#define circuit ZPUino_Vanilla
```

Figure 6-10 shows the sketch with the #define directive added; I renamed it *FirstDUO_1*.

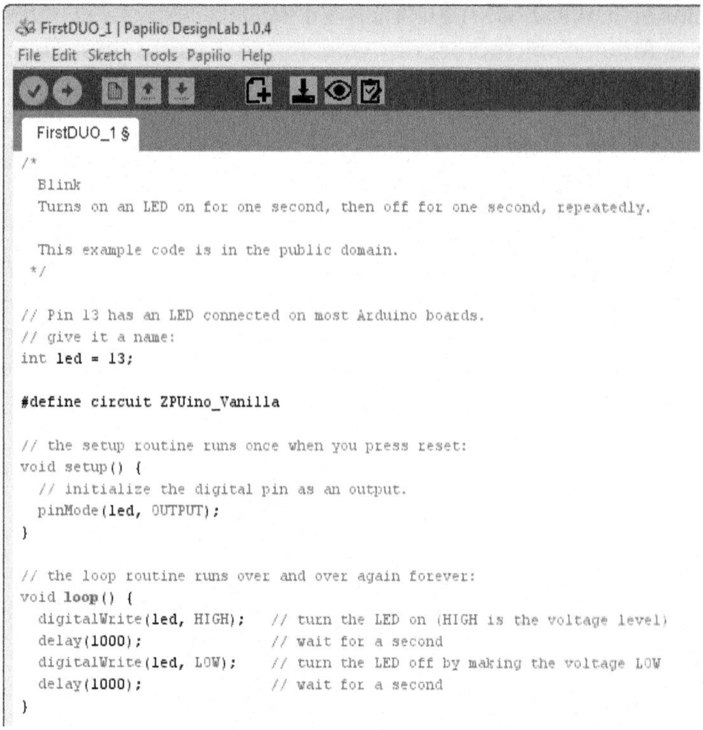

Figure 6-10 *Add circuit #define*

Step 5: Load FPGA Bit File

We're almost there. Now, load the FPGA bit file to the module by clicking the Load Circuit icon in the DesignLab toolbar (Figure 6-11).

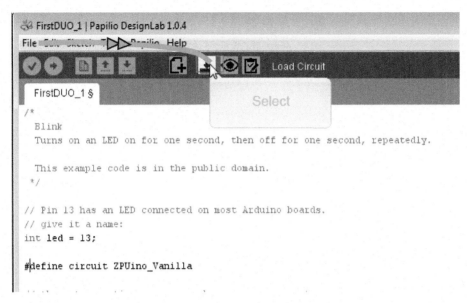

Figure 6-11 *Load circuit*

Wait for the "Done burning bit file" message to appear (Figure 6-12).

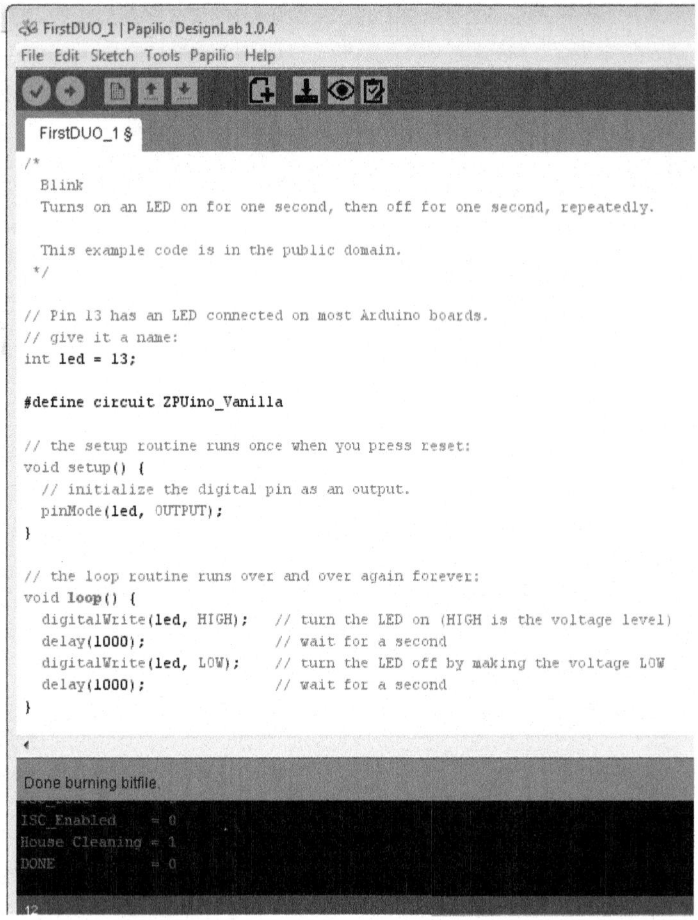

Figure 6-12 *Bit file done loading*

Step 6: Compile and Upload Sketch

The last step is to compile and upload the sketch (C program) to the ZPUino (Arduino soft core) so it can begin executing. Do this by clicking the Upload icon in the DesignLab toolbar, as seen in Figure 6-13. This is the same as if you were using the standard Arduino IDE.

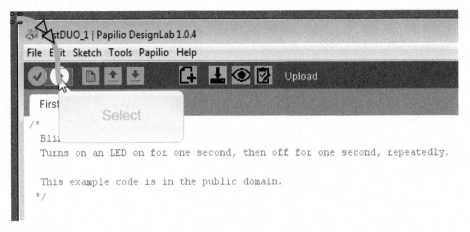

Figure 6-13 *Upload your sketch to the ZPUino soft core*

Wait for the "Done uploading" message, and you should see the green LED blinking on your board.

Getting Started with the DesignLab Video-Audio Player

Now that you have installed DesignLab and set up your DUO hardware, you are ready to begin your first SoC design project. Here's what you'll need for this project:

- A Papilio DUO and the DUO Computing Shield (you can get them as a bundle from the Gadget Factory website (*http://bit.ly/1VE5Cvd*))
- A display with a VGA input
- A small computer speaker with an audio jack
- The latest versions of DesignLab and Xilinx ISE WebPACK

How It Works

You can think of this SoC design as a mini PC inside the FPGA. Take a look at the block diagram in Figure 6-14.

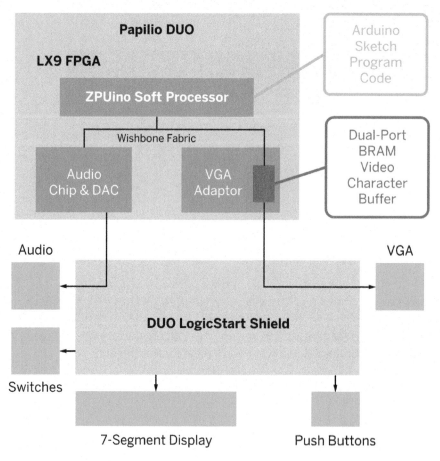

Figure 6-14 *Video-audio player block diagram*

 FYI

Note that there is a separate DAC device on the board.

The Papilio DUO contains our Xilinx LX9 FPGA; obviously, that's where our SoC will be going. The video-audio player design includes a ZPUino soft processor core. The ZPUino core was created by Alvaro Lopes; it is a 32-bit processor that is easily programmed, like the Arduino microcontroller we all know and love. You can read more about ZPUino at the ZPUino website (*http://www.alvie.com/zpuino/*). The ZPUino is connected to a Wishbone fabric (see "SoC Architecture") that provides us with the address, data, and control buses for our embedded system. We will be using a static memory map in this design, and access to our two peripheral devices (audio and VGA) will be through register I/O transactions.

Video data will be generated by the ZPUino soft processor core and written to the VGA adapter core. You can think of this as a PCI VGA adapter card in a PC. In the same way video data is written to a VGA peripheral card through a dual-port RAM buffer over the PCI bus, our VGA adapter contains a dual-port RAM that is made out of FPGA block RAM. The buffer is written by the soft processor over the Wishbone bus on one side of the dual-port RAM, and the VGA adapter takes the data out of the dual-port RAM on the other side of the buffer to convert it to VGA video signals. The VGA adapter takes care of all the high-speed signaling that is required by the VGA standard, just like a PCI VGA adapter card would do on a PC.

Similarly, for our audio we will be using a couple of audio cores, like the Yamaha YM2149 Software-Controlled Sound Generator (SSG) chip (*http://bit.ly/1QhwbT8*) that was used in many vintage arcade games (you can learn more about it at Wikipedia (*http://bit.ly/1VEaYXe*) and on the Gadget Factory website (*http://bit.ly/1VEaZKQ*)) and the SID (*http://bit.ly/1VEb3di*) (short for Sound Interface Device), which was used in the Commodore 64.

This highlights the advancement of technology—at one time there were specific chips to do specific functions, but now these functions are available in HDL code and take up relatively few resources in a small FPGA.

We will be connecting the VGA adapter and audio cores to our Wishbone bus, and programs running on the ZPUino soft processor core will be able to configure and access these cores through read/write register accesses. In other words, the soft processor will be writing data and control words to this device much like you do in a PC to a sound card. The output of the audio cores will be connected to a DAC (digital-to-analog converter) before it goes out to the I/O pins. Most computer audio signals are stored in digital form (for example, MP3s and CDs), and in order to be heard through speakers they must be converted to analog signals. DACs are therefore found in CD players, digital music players, and PC sound cards.

We'll be writing programs to execute on the soft processor core in Arduino C. Here, we can write directly into registers in the audio and VGA cores that will produce video effects on our screen and sounds from our speakers, which we'll connect to the logic start shield. We can also use library functions that take a lot of the work out of coding video and sound objects.

Design

We will be using DesignLab to design our video-audio player SoC, so go ahead and open DesignLab now. Follow these steps, and you'll be an SoC designer in no time!

Step 1: Create New DesignLab Project

1. Open a new ZPUino project by going to File→New ZPuino SOC Project.
2. Save the project into a new sketch by selecting File→Save As.

3. Give the project a new name. I called mine *Video_Audio_Player_SoC1*.

4. To open the Xilinx ISE, in the top toolbar, click Edit Circuit, then click Yes to copy the circuit to your local project (Figure 6-15).

Figure 6-15 *DesignLab Edit Circuit*

Step 2: Edit Your Design in Xilinx ISE

Now we need to edit the design:

1. In the Design→Implementation→Hierarchy view in ISE, double-click "Papilio_DUO_LX9 (Papilio_DUO_LX9.sch)," as shown in Figure 6-16. Here's where the magic begins. DesignLab has created your top-level schematic starting point for you. This top-level schematic is really a functional block diagram of your design. Each symbol has full HDL under it describing the design of the IP block it represents.

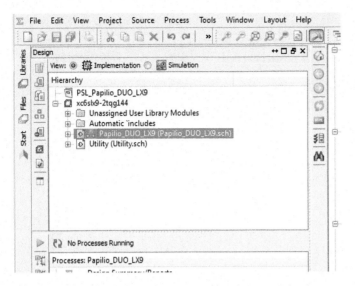

Figure 6-16 *Open the schematic*

2. Next, we need to delete the Papilio DUO pinout block and replace it with the Papilio DUO Computing Shield pinout block. Be sure to delete all the GPIO blocks too, as you won't be needing those. You can find the symbol for the Computing Shield in the *DesignLab\libraries\Papilio_Hardware>* library when you click the Add Symbol icon in the left toolbar. Figure 6-17 shows the starting schematic.

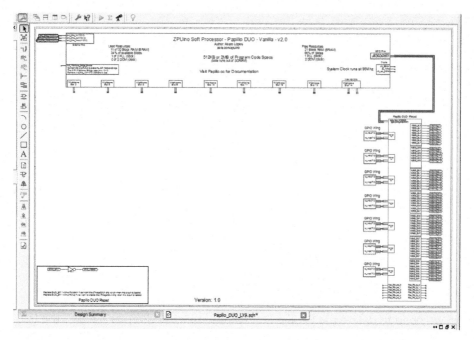

Figure 6-17 *Beginning schematic with DUO pinout block*

And Figure 6-18 shows the schematic with the Computing Shield pinout block.

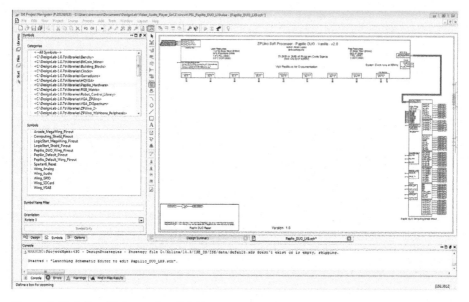

Figure 6-18 *Edits with Computing Shield pinout block*

Step 3: Add VGA Adapter Block

Next, we will add our VGA adapter IP block. Simply click the Add Symbol tool in ISE, and select the *DesignLab\libraries\VGA_ZPUino>* library from the Categories list and *VGA_ZPUino* from the Symbols list. Add a wire from the VGA adapter to Wishbone slot 14 of the ZPUino. Then connect the VGA_Bus of the VGA adapter to the VGA_Bus of the DUO Computing Shield pinout block, as seen in Figure 6-19. There's no need to worry about plugging into the right Wishbone slot, because each DesignLab IP block has a slot auto-sensing/detection function built into it. This saves you all the issues of manually sorting out the address map in the code. The autodetection feature also keeps track of multiple instances of the same type with an autolock feature. Nice job, Gadget Factory!

FYI

The Add Symbol tool is the first logic gate icon in the toolbar on the left.

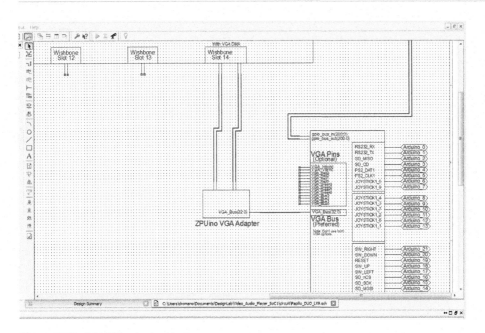

Figure 6-19 *Add VGA adapter block*

Step 4: Add Audio Blocks

We'll finish off our SoC design by adding a couple of audio generator blocks:

1. Select the *DesignLab\libraries\ZPUino_Wishbone_Periperals>* library from the Categories list.

2. Add the *AUDIO_zpuino_wb_YM2149* and *AUDIO_zpuino_wb_sid6581* symbols.

3. Connect the Wishbone bus of the YM2149 audio chip to Wishbone slot 6 of the ZPuino soft processor and the C64 SID audio chip to slot 8. Don't forget to connect the ck_1MHz clock of the SID to the ck_1MHz clock of the ZPUino, as seen in Figure 6-20.

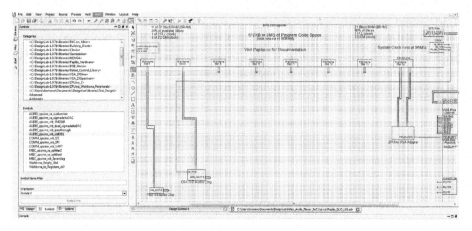

Figure 6-20 *Add audio chips*

4. Add *AUDIO_zpuino_sa_audiomixer* and *Splitter4* symbols and connect them to the output of the YM2149 and C64 SID. Also connect the ck (clock) of the audiomixer to the ck_96MHz clock of the ZPUino, as shown in Figure 6-21.

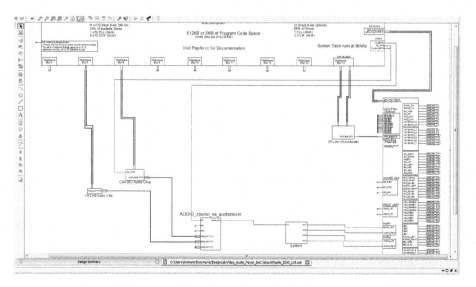

Figure 6-21 *Add audiomixer and splitter*

5. Lastly, just add a GND and a VCC symbol from the General category and connect the `rst` pin of the audiomixer to GND and the `ena` pin to VCC (Figure 6-22).

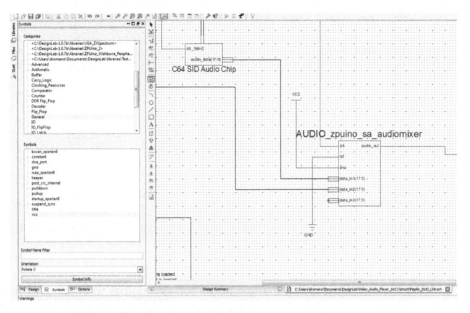

Figure 6-22 *Connect audiomixer control pins*

 FYI

If you want to see the HDL code of an IP block while you are editing the schematic, right-click the schematic symbol for the block whose source code you want to see and select "Push into symbol." Once you do that, you will be editing the source code for that symbol.

Step 5: Implement and Generate Bit File

Congratulations—you just completed your first SoC design, and you didn't need to write one line of HDL code. Your top-level schematic should look like Figure 6-23.

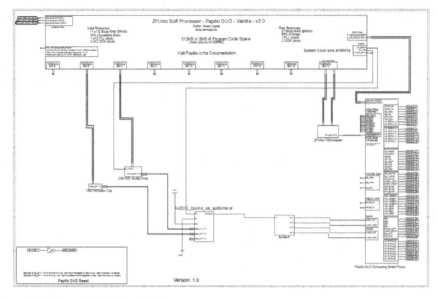

Figure 6-23 *Complete top-level SoC*

Save your design and click the green triangle in the top ISE toolbar to implement (build) your design. Then, in the Processes window, double-click "Generate Programming File."

Step 6: Create Sketch, Load, and Run

We'll need a sketch (program) to run on our ZPUino soft processor core, so we'll copy the example code from the Video_Audio_Player example into our DesignLab window:

1. Delete all the code for your SoC design from your DesignLab window.
2. Go to File→Examples→Video_Audio_Player.
3. Copy and paste all the code from the example into your design.
4. Save your design.
5. Connect your Papilio DUO with the Computing Shield installed to your PC using the USB-Mini connector.
6. Connect your VGA monitor and audio speaker to the Computing Shield.
7. Select your board: Tools→Board→Papilio FPGA Boards→Papilio DUO FPGA 2 MB (or 512 KB).
8. Select your COM port: Tools→Port→COMx (Papilio DUO FPGA).
9. Click Load Circuit in the DesignLab toolbar.
10. Click Upload and you should see the test menu on your VGA display, as shown in Figure 6-24.

Figure 6-24 *VGA audio player menu*

11. Use the push button on your Computing Shield to navigate the menu and make selections among the test files.

Experiments

Example 6-1 is a sketch you can use to experiment with writing to your VGA display. The example shows how you can write a color to a certain pixel using the library, and then how you can write a color to another pixel by writing directly to the framebuffer that is created in SRAM memory.

 FYI

If you try to read the location of the framebuffer directly from a register located in the hardware, which would be register0 on Wishbone slot 14, you will find that that register is write-only. You'll need to resort to reading the framebuffer memory location from the library. During initialization the library allocates the framebuffer in SRAM memory and then writes the base location of the allocated memory to register0 on Wishbone slot 14. The VGA controller then uses that shared memory to drive the VGA control lines to create the VGA picture. So, if you write directly to the framebuffer in memory, you will directly affect the output to the VGA monitor.

Example 6-1 *Writing a pixel directly to VGA*

```
   Gadget Factory - Simple example of writing a pixel directly to VGA Framebuffer or
using the library.
   Use this as a template for DesignLab ZPUino System on Chip Projects.
   To learn more about using DesignLab please visit http://learn.gadgetfactory.net.

   Tutorials:

   Related library documentation:

   Hardware:

   Special Notes:

   created 2015
   by Jack Gassett
   http://www.gadgetfactory.net

   This example code is in the public domain.
   */

#define circuit Computing_Shield2

#include <Adafruit_GFX.h>
#include <ZPUino_GFX.h>
#include <PLL.h>

// Assign human-readable names to some common 16-bit color values:
#define BLACK    0x0000
#define BLUE     0x001F
#define RED      0xF800
#define GREEN    0x07E0
#define CYAN     0x07FF
#define MAGENTA 0xF81F
#define YELLOW  0xFFE0
#define WHITE    0xFFFF

ZPUino_GFX gfx;

void setup() {
  // put your setup code here, to run once:

    Serial.begin(115200);
    delay(3000);

    gfx.begin( &modeline_640x384_60 );  //Highest mode supported by Papilio DUO 512KB

    Serial.println("Framebuffer Base Address in SRAM:");
    Serial.println(uint16_t(gfx.getFramebuffer()),HEX);

    //Write a RED pixel directly to the first address of the Framebuffer in SRAM
```

```
        gfx.getFramebuffer()[0] = RED;

        //Use the library to more easily write green to the first pixel of the tenth line
        //on the display
        gfx.setPixel(0,10,GREEN);

    }

    void loop() {

    }
```

Source Code

All the code is open source and can be found at DesignLab's GitHub (*http://bit.ly/1NU8uil*).

Takeaways

Here are some of the key takeaways from the exercises in this chapter:

- A hybrid hierarchical approach to the SoC design task—using schematic entry for the top level and HDL on the lower levels—can make the job much easier for novice SoC designers.

- Gadget Factory's DesignLab is a great frontend tool for Xilinx ISE schematic entry.

- The ZPUino soft processor core, created by Alvaro Lopes, is a 32-bit processor that is easily programmed like the Arduino and is a great building block for FPGA SoCs.

- DesignLab provides a full integrated solution for designing and programming SoCs using ZPUino.

Just for the Fun of It | 7

Old Arcade Games Made New with FPGAs

It seems that there is a universal appeal to vintage arcade games like Pac-Man and Space Invaders. At a recent Maker Faire, I had a demo of the FPGA arcade projects we will be completing in this chapter running in my booth. I was amazed that people, young and old alike, all seemed to know Pac-Man and were instantly compelled to play it. Let's face it; these old games are just plain fun!

There are some very good PC-based arcade game emulators out there that give you the ability to play these vintage games—but playing them on a laptop is just not the same as it was on an Atari. With FPGAs, though, you can re-create the real look and feel of the original games! In this chapter, I'm going to show you how easy it is, thanks to our friends at Gadget Factory and the Papilio DUO.

I'll be showing you two cool projects in this chapter that are variations on the vintage arcade theme. One uses a traditional VGA display and the other uses a really cool LED dot matrix display. So what are we waiting for? Let the games begin!

 FYI

I'll be using the Papilio DUO exclusively for these projects, but with a little work you can remap these designs to other FPGA boards. You may need to create some additional breadboard circuits to get the I/O you need, though, which the Papilio DUO Computing Shield provides (this topic is beyond the scope of this book).

Getting Started with VGA-Displayed Arcade Games

Here's what you'll need for our first project:

- A Papilio DUO and the DUO Computing Shield. You can get them as a bundle (*http://bit.ly/1VE5Cvd*) from the Gadget Factory website.

- One or two vintage Atari game controllers. You can find them online (*http://bit.ly/ 1VE5Diy*), or you can try to make one of your own (*http://bit.ly/1VE5DPD*). I like using the original Atari 2600 controllers—they just add to the vintage experience.

- A small LCD display with a VGA input. You'll need to physically turn this display 90 degrees to view the game. A VGA cable is also required. I held the display up using a small artist's table easel. The setup is shown in Figure 7-1.

Figure 7-1 *VGA arcade project setup*

Thanks to our friends at Gadget Factory, we can jump right in with playing games using our Papilio DUO and the DUO Computing Shield hardware. But first, you'll need to understand some technical details.

How It Works

There are a couple of key design features you will need to understand before we get into loading and playing our vintage arcade games. If you check out the diagram in Figure 7-2, you will see that our two hardware modules, the Papilio DUO and the DUO Computing Shield, connect together. The Papilio DUO contains our FPGA, and Computing Shield is our I/O module, where we will connect our VGA display, audio, and joysticks.

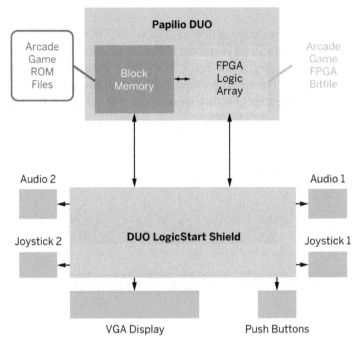

Figure 7-2 *Papilio DUO arcade system*

There are two design elements that are required for the games to operate. The first is our usual FPGA design bit file. This is the file that contains our FPGA hardware logic design. The other element is the actual arcade game ROM files. This is the part that gets a little tricky, because I won't be directly providing you with any information on where and how to get the ROM files. Most of these ROMs are still under copyright. Let's just say that ROM files for vintage arcade games exist on the Web. I will explain how you can use a ROM file on the Papilio DUO system if you have one. The design bit files and design source code are provided by Gadget Factory; these are all open source and free to use.

 FYI

Gadget Factory does provide you with a couple of ROM images that are open source, which you can experiment with.

Loading a Game

You'll be playing some of your favorite arcade games in no time just by following these few easy steps.

Step 1: Installing ROMVault

We start by using Gadget Factory's very cool version of ROMVault. ROMVault (*http://www.romvault.com*) is a tool that was created as a Windows utility for verifying and managing ROMs, like those used in vintage arcade games. You can visit the website to find out more about it if you like; for the Papilio, we are interested only in the RomVault-Papilio-Edition that you can download from the Gadget Factory website (*http://bit.ly/1JWs9lA*) (see Figure 7-3).

Figure 7-3 *RomVault-Papilio-Edition download*

Once you download the ZIP file you will need to create a directory off of your C drive to extract it into.

Heads Up!

Be sure to create a directory right off of your C drive, and do not use any spaces in the directory name!

Your directory setup should look something like Figure 7-4 when you are done.

ROMVault
  RomVault-Papilio-Edition-2.2.1
    datroot
    images
    library
    papilio
    romroot
    Unsorted

Figure 7-4 *Example directory structure for RomVault-Papilio-Edition*

Heads Up!

If you get any unexpected copying errors during the extraction process, like the one in **Figure 7-5***, just ignore them by checking "Do this for all items" and clicking Skip. These errors have no impact on the function of the tool.*

Figure 7-5 *Copying error*

Step 2: Running ROMVault

Now you just need to double-click the ROMVault application in the top directory, as seen in Figure 7-6, to get it started.

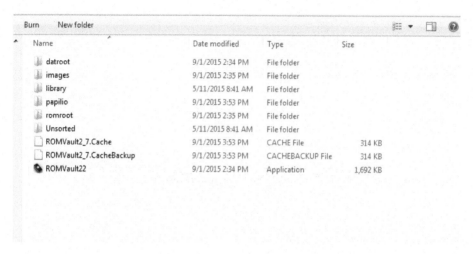

Figure 7-6 *Launch the ROMVault app*

Now you should be looking at the RomVault-Papilio-Edition screen (Figure 7-7).

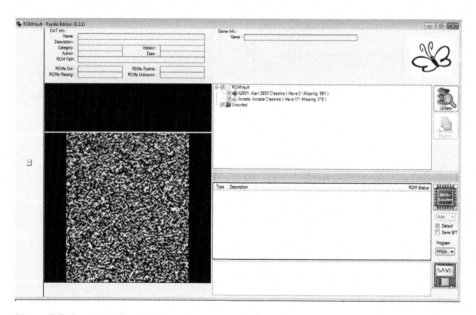

Figure 7-7 *RomVault-Papilio-Edition*

Click "Arcade: Arcade Classics" and then scroll down in the ROM status window, and you should see Mr. Do's Nightmare and Pong Demo highlighted in green. These two games have ROM files included in the download.

Step 3: Programming the FPGA

Connect your Papilio DUO system to your computer's USB port, connect your VGA monitor to the VGA port of the DUO Computing Shield, and plug your joystick into Joystick port 1 on the computing shield.

Double-click the Pong Demo, and RomVault-Papilio-Edition does the rest; it loads the bit file to the FPGA and maps the ROM files. After a few seconds you should see the game on your VGA display—it's now ready to play. Have fun!

Heads Up!

These old arcade games were designed to run on VGA displays that were turned 90 degrees and mounted in game console cabinets, so don't get alarmed when you see your game displayed sideways. You will need to turn your monitor to match the original physical screen orientation.

Source Code and ROM Files

You can find the FPGA source code for the Papilio DUO arcade designs on GitHub (*http://bit.ly/1VE5Li4*).

Keep in mind that the ROM files are handled differently in RomVault-Papilio-Edition than they would be if you were to try to use them directly in the Xilinx ISE WebPACK. In ISE, the ROM files are mapped directly to logic gates using the *romgen* tool, whereas in RomVault-Papilio-Edition they are mapped to BRAM. This is done so you don't have to generate a new bit file every time you load new ROM files. Figure 7-8 shows the Pac-Man game source code hierarchy in ISE.

Figure 7-8 *Pac-Man source code hierarchy*

I won't be getting into how to modify the VHDL code for these games here, but you are free to experiment with it. Remember, it's all reprogrammable, so you have nothing to lose.

RomVault-Papilio-Edition does provide you with a handy method for checking if any ROM files that you may come across will work with the Papilio hardware. For example, if you click Pac-Man in the RomVault-Papilio-Edition ROM Status window you will get some good information about the ROM, as seen in Figure 7-9.

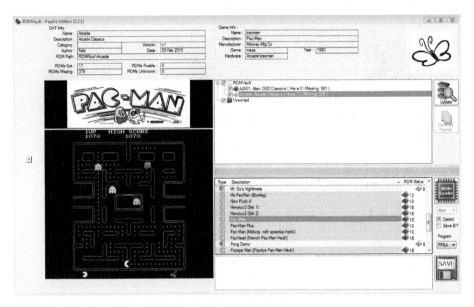

Figure 7-9 *RomVault-Papilio-Edition Pac-Man*

Notice that the Game Info section provides you with the filename of the ROM; in this case, it is "pacman." You also can see in the DAT Info section that the ROM path is *ROMRoot \Arcade*. This is the directory where you will put any ROM ZIP files you happen to come across. Once you do that, then all you need to do is double-click the game in the ROM Status window, and you should be good to go.

FYI

*If you change the selection in the Program drop-down from FPGA to Flash, as shown in **Figure 7-10**, you will load the game to the flash memory of the DUO, so you won't lose the game on a power cycle.*

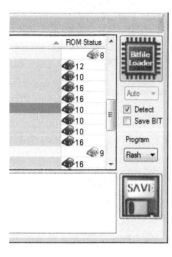

Figure 7-10 *Load from Flash*

Getting Started with LED Dot Matrix–Displayed Arcade Games

This next project is just plain cool! We are going to run a Pac-Man–like game on an LED dot matrix display. Here's what you need to get started:

- A Papilio DUO, available from the Gadget Factory website (*http://bit.ly/1TyJF1j*)

- A Papilio RGB LED Panel Wing, available from the Gadget Factory website (*http://bit.ly/1VE5QCn*)

- A Papilio Platform Joystick Wing, available from the Gadget Factory website (*http://bit.ly/1VE5RpQ*)

- A 2.54 mm pitch 16 pin female-to-female IDC ribbon cable (Newegg item number 9SIA4SR1PN3714 (*http://bit.ly/1OZEO9w*) or similar)

- A 32 × 32 RGB LED matrix panel, 6 mm pitch, available from Adafruit (*http://bit.ly/1VE5Sdt*) (Adafruit provides a lot of information about this display technology on the product page)

- A 5V 2A (2000 mA) switching power supply, available from Adafruit (*http://bit.ly/1VE5Pyo*)

- A female DC power adapter (2.1 mm jack to screw terminal block), available from Adafruit (*http://bit.ly/1VE5Tyc*)

- One or two vintage Atari game controllers (you can find these online at DKOldies (*http://bit.ly/1VE5Diy*))

- The latest versions of DesignLab and Xilinx ISE WebPACK

Figure 7-11 shows what the setup for this project looks like.

Figure 7-11 *LED dot matrix project setup*

How It Works

This design is basically another SoC design from the inventory of design examples provided by Gadget Factory's DesignLab. It's based on the 1-Pixel Pac-Man project by Mike Szczys, which you can read about on hackaday (*http://bit.ly/1VE5Mm6*).

Gadget Factory's FPGA version is called Matrixman, and all the code you need to build it comes with the latest version of DesignLab. The Matrixman design is similar to our Video Audio Player SoC design from Chapter 6. It also uses the ZPUino soft processor core and the Wishbone bus fabric, only this time we are using an RGB adapter block instead of a VGA adapter to generate our display (see Figure 7-12).

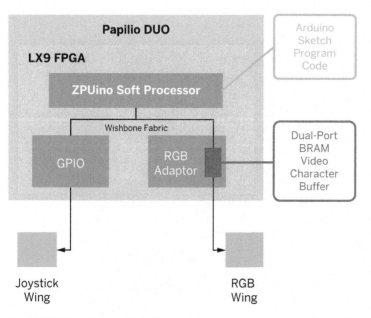

Figure 7-12 *Matrixman block diagram*

Memory Mapping

When we think about computers, most of us think of the PC as the prime example. We may even say, "I'm going to work on my computer," referring to our PC. In reality, the personal computer (PC) is just one class of computing device. A PC is what is commonly referred to in technical terms as a *general-purpose computing device*. A lot of engineering has gone into PC architecture over the years to make it just such a device. Most of this effort is under the covers and, to most users, way below the radar. Historically, early in the evolution of the PC, it became clear that consumers wanted to be able to open their computers and add expansion cards to customize or change the functionality—just like with the rack system we talked about in Chapter 6. This caused some serious headaches in the early days, and some of you may even remember the notorious "bus conflict" or "IRQ error"... ahhh!

This whole mess came about because of how the CPU cores of the early PCs understood where devices like graphics cards, hard drives, floppy drives, and sound cards connected to them lived in the system. This in turn determined how the CPU communicated with these devices.

As an example, think of the street you live on. Every house on the street has a fixed address. If you wish to communicate with your neighbor by mailing a letter, you first need to write the letter and then send it through the mail system by putting your message in an envelope and writing your neighbor's name and address on the front of the envelope. You then put the letter in the mailbox, and the postal service retrieves the letter, processes it, and delivers it to the person whose name and address appear on the front of the envelope.

Now suppose there is an empty lot on your street, and a home developer comes in and builds a new house—but the addition of this new house changes the addresses of all the houses on the entire street, unbeknownst to you and to the mail system. Now what happens when you mail a letter to Joe at 114 ISA Drive, but that's not Joe's address anymore? It's Bill's, so

Bill now starts getting Joe's mail and doesn't know what to do with it. He calls the post office and screams, "Bus conflict error!" You get the picture.

PC engineers needed a clever way to change the static addressing method used by the CPU into a dynamic method that could accommodate moves, additions, and changes in devices connected to the CPU bus—and that's why the Peripheral Component Interconnect (PCI) standard came into being. PCI is a very complex standard that goes way beyond the scope of this book, but it's important to us as SoC design wannabes to understand the differences between a static and a dynamic memory mapping system.

There is another class of computing device known as *embedded computing devices*. An embedded computing device is a computer purchased as part of some other piece of equipment. Typically, it contains dedicated software (which may or may not be user customizable). It often replaces electromechanical components and has no "real" keyboard and a limited display, or no general-purpose display. Examples of embedded computers are all around us: they are in our cars and the airlines we fly on; they are in our home appliances, cable boxes, WiFi routers, video game boxes, cell phones, etc. It's hard to touch anything in our world today that doesn't have an embedded computer in it. Our smartphones and tablets are really an extension of embedded computing. When was the last time that you opened up your smartphone and put in a new graphics card?

Embedded computers have mostly closed, dedicated design architectures and therefore are free from the PC's dynamic memory map dilemma and all the extra baggage of the PCI standard, MS Windows bus enumeration, and IA architecture configuration complexities. Most SoCs that contain a CPU are designed for embedded systems and follow a simple static memory map architecture. In other words, the address locations for devices in the system are fixed at design time and are not changed. We will be focusing on simple fixed memory-mapped systems like this design, which may be considered embedded designs.

Design

To get started with Matrixman, just follow these few easy steps; DesignLab does the rest.

Step 1: Open Example Design

1. Open DesignLab and go to File→Examples→*SmartMatrix_32x32/matrixman*. Your screen should look like Figure 7-13.

Figure 7-13 *Matrixman in DesignLab*

2. To view your design, click View Circuit in the top toolbar. A PDF file of the top-level schematic will open, as seen in Figure 7-14.

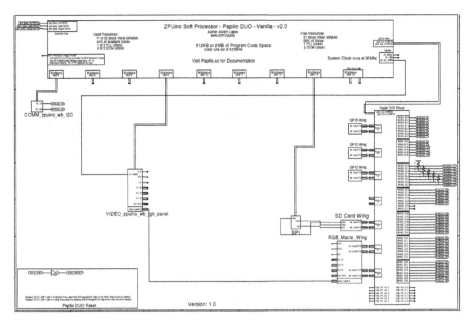

Figure 7-14 *Matrixman circuit view*

3. Connect your Papilio DUO with the RGB wing and joystick wing installed to your PC, using the USB-Mini connector.

4. Connect your LED matrix and joystick.

5. Select your board: Tools→Board→Papilio FPGA Boards→Papilio DUO FPGA 2 MB (or 512 KB).

6. Select your COM port: Tools→Port→COMx (Papilio DUO FPGA).

7. Click Load Circuit in the DesignLab toolbar.

8. Click Upload, and you should see the Pac-Man game ready to play!

9. Have fun!

Experiments

Here are a couple of examples of sketches you can use to experiment with writing to your LED dot matrix display.

You can write directly to the hardware by running a sketch like the one in Example 7-1 (it will turn the first pixel white).

Example 7-1 *Direct pixel write*

```
*/
#define circuit RGB_Matrix
void setup() {
```

```
    // put your setup code here, to run once:

        REGISTER(IO_SLOT(9),0x1000) = 0xFFFFFF;
}

void loop() {

}
```

To turn the first pixel red using the library, use the sketch in Example 7-2.

Example 7-2 *Library pixel write*

```
#define circuit RGB_Matrix
#include <SmartMatrix_32x32.h>

SmartMatrix matrix;

const rgb24 black = rgb24(0x0, 0x0, 0x0);
const rgb24 white = rgb24(0xff, 0xff, 0xff);
const rgb24 red = rgb24(0xff, 0x0, 0x0);
const int defaultBrightness = 15*(255/100);

void setup() {
  // put your setup code here, to run once:

    Serial.begin(115200);
    delay(3000);

    matrix.begin();
    matrix.setBrightness(defaultBrightness);

    matrix.setColorCorrection(cc24);

    matrix.fillScreen(black);

    matrix.drawPixel(0,0,red);

    matrix.apply();

}

void loop() {

}
```

Source Code

All the code is open source and can be found on GitHub (*http://bit.ly/1NOjHTZ*).

Takeaways

Here are some of the key takeaways from the projects in this chapter:

- The Papilio DUO can be used to easily build cool projects that are variations on the vintage arcade theme. One of our projects used a traditional VGA display, and the other used an LED dot matrix display.
- You can build on the Matrixman example through programming a ZPUino sketch.

Cha-Ching! 8

Bitcoin Mining Project

The California Gold Rush began in 1849 when gold was discovered in the Sacramento Valley. It wasn't long before news spread that there was *gold in them there hills*, and thousands of prospective gold miners traveled to San Francisco and the surrounding area by land and sea. By the end of 1849, the population of the California territory had grown from less than a thousand to over 100,000. In 1852, the Gold Rush peaked; by then, 2 billion dollars' worth of gold had been extracted from the territory.

You may be asking, what in the world does this have to do with FPGAs? In a sense, some believe that a new Gold Rush is on in the form of Bitcoin currency, and with the help of FPGA technology, you can become a Bitcoin miner.

You can learn all about Bitcoin from Bitcoin's FAQ (*https://bitcoin.org/en/faq*) and the Bitcoin Mining website (*https://www.bitcoinmining.com*), which describes it as follows:

> Where do bitcoins come from? With paper money, a government decides when to print and distribute money. Bitcoin doesn't have a central government.

> With Bitcoin, miners use special software to solve math problems and are issued a certain number of bitcoins in exchange. This provides a smart way to issue the currency and also creates an incentive for more people to mine.

According to the Bitcoin FAQs:

> Bitcoin is a consensus network that enables a new payment system and a completely digital money. It is the first decentralized peer-to-peer payment network that is powered by its users with no central authority or middlemen. From a user perspective, Bitcoin is pretty much like cash for the Internet. Bitcoin can also be seen as the most prominent triple entry bookkeeping system in existence...Bitcoin is the first implementation of a concept called "cryptocurrency," which was first described in 1998 by Wei Dai on the cypherpunks mailing list, suggesting the idea of a new form of money that uses cryptography to control its creation and transactions, rather than a central authority.

In this chapter, we'll learn how to become Bitcoin miners through an easy and fun FPGA SoC project.

FYI

Bitcoin mining technology has moved so rapidly that FPGA mining is no longer an effective way to find Bitcoins. A part of Bitcoin technology is that the difficulty of the mining algorithm is increased as new methods are created. When the first ASIC rigs were created, they increased the difficulty to the point where FPGAs can no longer find blocks on their own. So, unless you join a mining pool, it is very improbable that you will find anything, and even with a pool it can take quite some time, if it happens at all. Overall, Bitcoin mining on an FPGA is really just a fun exercise at this point; it is not a practical way to find Bitcoins.

Once again we'll be turning to our friends at Gadget Factory and DesignLab to help us quickly and easily build a Bitcoin SoC.

FYI

I'll be using the Papilio DUO and DesignLab exclusively for this project, but as always, with a little work you can remap this design to other FPGA boards. You may need to create some additional breadboard circuits to get the I/O you need, though, which the Papilio LogicStart Shield provides. This again is beyond the scope of this book.

Getting Started with the Bitcoin Miner

Here's what you'll need for this project:

- A Papilio DUO and the LogicStart Shield; you can get these from the Gadget Factory website (*http://store.gadgetfactory.net*)
- The latest versions of DesignLab and Xilinx ISE WebPACK

The setup is shown in Figure 8-1.

Figure 8-1 *Bitcoin setup*

Our picks and shovels for this project will be provided by Gadget Factory's DesignLab and the Papilio DUO. Before we begin, let's take a few moments to review our design.

How It Works

This design is actually quite simple. The block diagram in Figure 8-2 shows the main Bitcoin block connected to the USB controller that is included on the Papilio DUO board. Your PC is connected to the board through the USB-Mini connector. The PC communicates with the Bitcoin logic over a USB serial link. The Bitcoin block is also connected to the 7-segment display, LEDs, and push buttons of the Papilio LogicStart Shield.

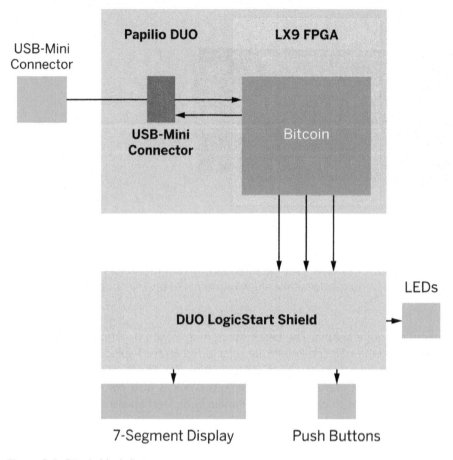

Figure 8-2 *Bitcoin block diagram*

The Bitcoin block contains all the crypto logic that is used to perform the hash algorithm in the Bitcoin mining operation. LED0 and LED1 are connected to the RX and TX lines of the Bitcoin block.

Design

Just follow these easy steps, and you'll be panning for Bitcoin gold:

1. Open DesignLab and go to File→Examples→BitCoin_Miner →BitCoin_Miner (see Figure 8-3).

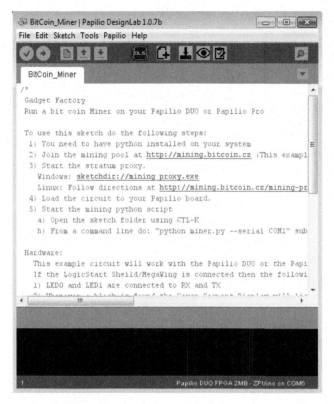

Figure 8-3 *DesignLab Bitcoin example sketch/design*

2. To view your design, click View Circuit in the top toolbar. A PDF file of the top-level schematic will open, as seen in Figure 8-4.

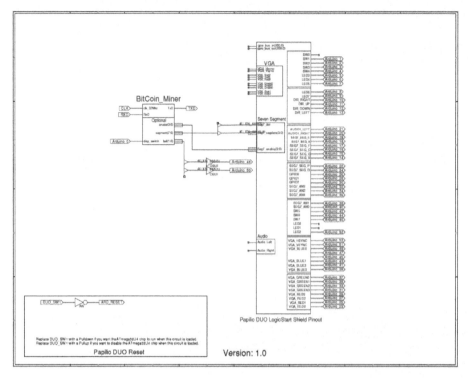

Figure 8-4 *Bitcoin DesignLab schematic*

3. Connect your Papilio DUO with the LogicStart Shield installed to your PC, using the USB-Mini connector.

4. You need to have Python (*https://www.python.org/downloads/*) installed on your system.

5. Join the mining pool (*http://mining.bitcoin.cz*). (The DesignLab example is set up to use a Gadget Factory test account, so you can skip this step if you just want to test it. But without joining a pool, you really don't have a chance of getting any hits.)

6. Start the Stratum proxy. On Windows, it's in the DesignLab install directory (*DesignLab/Libraries/BitCoin_Miner/examples/BitCoin_Miner/mining_proxy.exe*), or if you create a new project with a copy of the Bitcoin Miner example, it will be in the top directory of your new Bitcoin project.

   Linux users should follow these directions (*http://mining.bitcoin.cz/mining-proxy-howto*).

7. Select your board: Tools→Board→Papilio FPGA Boards→Papilio DUO FPGA 2 MB (or 512 KB).

8. Select your COM port: Tools→Port→COMx (Papilio DUO FPGA).

9. In the DesignLab toolbar, click Load Circuit.

10. Start the mining Python script:

    a. Open the sketch folder using Ctrl-K.

    b. On your command line, type **python miner.py --serial COMx**, substituting your COM port.

11. Whenever a block is found, the 7-segment display will light up. The pattern will change as new blocks are found. SW0 will turn off the display to conserve power (up is on, down is off).

Heads Up!

After you install Python, be sure to update your system path to include its location!

Source Code

All the code is open source and can be found on GitHub (*http://bit.ly/1Pzs1HP*).

Takeaways

Here are some of the key takeaways from the project in this chapter:

- Bitcoin is a consensus network that enables a new payment system and a completely digital currency.

- With Bitcoin, miners use special algorithms that can be implemented in FPGAs to solve math problems and are issued a certain number of Bitcoins in exchange.

- The Papilio DUO can be used to easily build a Bitcoin mining rig.

- Bitcoin mining technology has moved so rapidly that FPGA mining is no longer an effective way to find Bitcoins. Overall, Bitcoin mining on an FPGA is really just a fun exercise at this point.

I Hear You!   **9**

Software-Defined Radio (SDR) on FPGA

Software-defined radio (SDR) seems to have been the Holy Grail of embedded design for years. I remember running into this design challenge multiple times during my engineering career, and I would always get very pumped up about the possibilities, only to come to the conclusion that the technology wasn't quite ready for primetime.

There are a number of definitions of software-defined radio, also known as software radio. The SDR Forum, along with the Institute of Electrical and Electronics Engineers (IEEE), defines SDR as a "radio in which some or all of the physical layer functions are software defined." It seems that their definition is stating the obvious. In this case, a radio is any kind of device that wirelessly transmits or receives signals in the radio frequency (RF) part of the electromagnetic spectrum to transfer information. In other words, SDR is a radio communication system where components that have typically been implemented in hardware (e.g., mixers, filters, amplifiers, modulators/demodulators, detectors, etc.) are implemented through software, typically in an embedded system or PC. As I stated earlier, the concept of SDR is not new, but with the rapidly developing capabilities of FPGAs and digital electronics it has become practical and affordable to implement these processes, which used to be only theoretically possible.

Radios exist in most of our modern gadgets, such as cell phones, computers, car door openers, vehicles, televisions—the list goes on and on, so you can see the attractiveness of SDR. Can you image having a single programmable hardware module that could implement all these different radio systems in one device? You get the picture. That's still a ways away from a technology standpoint, but we can experiment with this technology in a fun and practical way.

In this chapter, we'll discover the world of SDR by experimenting with an AM radio receiver implemented on the Red Pitaya platform. So what are we waiting for? Let's tune in and hear who's out there!

Implementation Technologies

There are a few different options when it comes to manufacturing an SoC once it has been designed and validated.

Each of these implementation technologies has tradeoffs that can be evaluated based on design criteria, centered around three variables: density, performance, and cost.

Let's take a look at the relative characteristics of the different technologies:

Full custom
> Very high transistor density, optimum performance in terms of clock speed. Involves the creation of a completely new chip, which consists of about a dozen masks (for the photolithographic manufacturing process). The first chip is very expensive to make, but thereafter each one is cheaper. An Intel CPU chip is a good example of a full custom design.

Standard cell
> Less dense and lower performance than full custom design. The designer uses a library of standard cells with an automatic place and route tool doing the layout. Still involves creation of custom chip, so all masks must still be made; manufacturing costs the same as for full custom.

Gate array
> Transistor density and performance can be almost as good as with standard cell design, but the production cost is lower. The designer uses a library of standard cells. The design is mapped onto an array of transistors, which is already created on a wafer; wafers with transistor arrays can be created ahead of time.

FPGA
> Performance is usually several factors to an order of magnitude lower than with the standard cell approach. Densities are an order of magnitude lower than with standard cell but an order of magnitude higher than with CPLDs. Much higher device cost than with other approaches, but FPGAs are reusable.

> FPGAs have a high component case than Custom and Semicustom, but significantly lower upfront costs, NRE. (Got my Xilinx hat on.)

Complex programmable logic device (CPLD) or erasable programmable logic device (EPLD)
> Less dense than FPGAs; higher cost per gate but does not need to be reprogrammed after power down. Very low power. Performance similar to FPGAs.

Most SoCs are implemented in standard cell technology, with the exception of some very high-performance designs. For us, the FPGA is best because it is reprogrammable, allowing us to reuse the same chip for an infinite number of designs. The performance and density of the FPGAs that we are using in this book is more than sufficient for us to experiment with some cool IP blocks from OpenCores (*http://open cores.org*). We will not be building tens of thousands of copies of our designs, so the cost per chip is not something we are very concerned with.

Getting Started with the SDR Receiver

I'll be walking us through the SDR receiver tutorial (*http://bit.ly/1TyJCTc*) that is provided by Red Pitaya. You will need to go to the tutorial page for details and downloads to complete this project.

FYI

We'll be using the Red Pitaya platform for this project exclusively; it is beyond the scope of this book to explore other platform options, but feel free to consider it.

Here's what you'll need for this project:

- The Red Pitaya platform, which you can order from one of their distributors (*http:// redpitaya.com/about/*)
- A 4–8 GB micro-SD card for your Red Pitaya
- A USB-Mini cable (I also used an Ethernet cable to connect the Red Pitaya board to my network)
- Some type of antenna setup connected to the IN2 SMA connector of the Red Pitaya (I used some things I had lying around my lab, like an SMA male to SMA male plug RF pigtail coax jumper cable, alligator jumper wires, and an old portable antenna). You can try the one described on the SDR receiver page using a four-wire telephone cable. With my setup I was able to get a few AM stations.

Figure 9-1 shows my setup for this project.

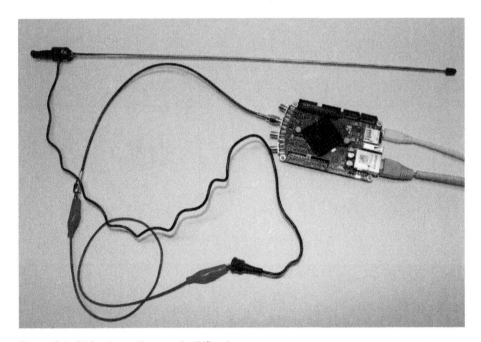

Figure 9-1 *SDR setup with my makeshift antenna*

How It Works

This project is by far the most technically complex one in this book. I'm presenting it to you just to provide you with a glimpse of what is possible with FPGAs and to give you a chance to have some fun with using the design as is. I will only be highlighting the design details here and will not be going into any technical depth. Believe me, we would need an entire book to cover what is going on in this design! I think it's just fun to play with the SDR application while exploring radio technology.

In the very high-level block diagram shown in Figure 9-2, we see the data flow model of our SDR. The antenna is connected to the RF frontend of the Red Pitaya and the analog-to-digital converter (ADC), which connects to the FPGA implementation of the digital down-converter (DDC). This is where the SDR magic happens; where all those ones and zeros coming in from the ADC are converted to digital symbols that can be interpreted by the SDR application running on the PC, which are transferred by the ARM core through the shared memory buffer, implemented in the BRAM of the Zynq programmable device.

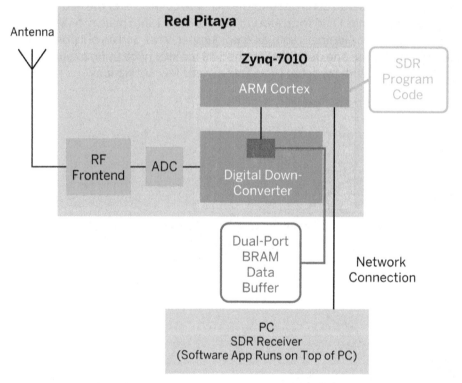

Figure 9-2 *Red Pitaya SDR block diagram*

According to the tutorial page (*http://bit.ly/1TyJCTc*), "The data coming from the ADC is processed by [an] in-phase/quadrature (I/Q) digital down-converter (DDC) running on the

Red Pitaya's FPGA." You can research what exactly the digital signal processing (DSP) elements do on the dspGuru website (*http://www.dspguru.com/dsp/articles*).

Heads Up!

You need more than just an FPGA to pull off this design. You need an RF (anolog) frontend and an ADC, which are electronic components outside of the FPGA device. You also need the ARM core. For the most part, this design is specific to the Red Pitaya board, which implements all the necessary circuitry. The principles learned here can be applied to other boards that have similar features.

Red Pitaya Setup

Heads Up!

Please note that due to the rapidly changing world of FPGA development boards and development technology, the setup procedures described here are subject to change. The following procedures were accurate at the time of writing, but may have changed since.

I like to rush right in, so the Quick Start (*http://staging1.redpitaya.com/quick-start/*) on the START dropdown menu on the Red Pitaya site looked good to me.

First, you need to choose a connection method that allows you to connect to the Internet. I chose the "Wireless connection" method using the suggested USB WiFi dongle (Edimax EW7811Un). I then simply followed the instructions provided by the online quick start guide. I was also able to get the seven-day evaluation of Red Pitaya's new Visual Programming tool running. This tool appears to use the MIT Scratch graphical block programming paradigm to write programs that run on the ARM core, sort of like Arduino sketches.

Loading the SDR

You'll be tuning in to AM radio in no time just by following these few easy steps.

Step 1: Copy Red Pitaya SD Card Image

1. You need to copy the SD card image for the Zynq that you get from the Red Pitaya SDR receiver tutorial page. Download the ZIP file and unzip it to a clean directory on your PC. Then just insert your SD card into your PC, delete any files that are on it, and copy the contents of the unzipped image to the SD card.

2. Once you do that you can plug the SD card into your Red Pitaya, connect your USB mini-cable to your PC, and power up the board.

Step 2: Install SDR Applications on PC

1. You next need to download the Windows SDR applications to your PC. You can find the links for them on the SDR receiver tutorial page. The programs are SDR# and HDSDR.

2. For the SDR# download, just unzip the file to a clean directory, and for HDSDR use the convenient Windows install executable.

3. Next, you will need to download the prebuilt ExtIO plug-in for SDR# and HDSDR. Then, copy *ExtIO_RedPitaya.dll* into the SDR# and HDSDR installation directories.

Step 3: Connect Red Pitaya to the Network

For this step, I just connected my Red Pitaya directly to my router with an Ethernet cable. I didn't try to use the wireless connection for this experiment. Once I'd connected my Red Pitaya to the network, I opened the administration page of my router and found its IP address by looking at the attached devices list. You may have to consult your router's user manual to find out how to do this for your model; typically you just need to type the IP address 192.168.1.1 into your browser and you should get your router's administration login page. You'll need to know the username and password for your router. The default setting for many routers is something like username = *admin*, password = *password*. Once you log in to your router, you will need to find the attached devices list; from there, you should see the Red Pitaya listed under "wired devices."

Step 4: Run SDR Applications

1. Start either the SDR# or the HDSDR application.

2. To run SDR#, go to the directory where you unzipped the file and double-click SDRharp (see Figure 9-3).

Figure 9-3 *SDRSharp program*

To run HDSDR, go to Start→All Programs→HDSDR→HDSDR.

3. Select Red Pitaya SDR from the Source list in SDR# or from the Options (F7)→Select Input menu in HDSDR.

4. Click the Configure icon in SDR# or the ExtIO button in HDSDR, then type in the IP address of the Red Pitaya board and close the configuration window.

5. Click the Play icon in SDR# or the Start (F2) button in HDSDR.

Figure 9-4 shows the SDR# application window.

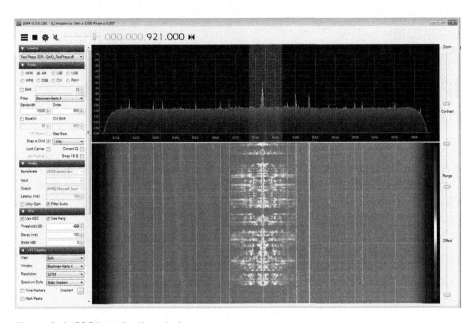

Figure 9-4 *SDR# application window*

I like the SDR# interface because I can see the peaks in the spectrum; when you click a peak, you typically find an active station.

The HDSDR application window is shown in Figure 9-5.

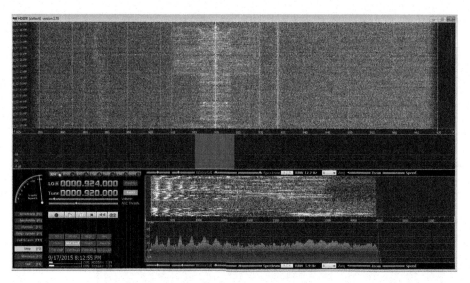

Figure 9-5 *HDSDR application window*

Source Code

All the code is open source, and links are provided on the Red Pitaya SDR receiver page (*http://bit.ly/1TyJCTc*). You will need to build a whole Linux Ubuntu development environment to work on the code for this design, and you may also need to build the new Xilinx Vivado FPGA environment. The Red Pitaya tutorial provides some details on how to do this; I will not be going into that here.

Xilinx ISE WebPACK Development

According to the engineers at Red Pitaya, from Red Pitaya release v0.94 (the latest), all the RTL files can be compiled using ISE except for one that is written in SystemVerilog (available at http://bit.ly/1mo7KMa).

*It's important to note that all the test bench files are written in SystemVerilog, which is a more powerful verification language. Neither the ISE nor the Vivado built-in simulators support SystemVerilog, so Red Pitaya uses the free version of ModelSim (**http://bit.ly/22ZTZUm**) provided by Altera for simulations.*

*In release v0.93 (**http://bit.ly/1TyJAuy**) of Red Pitaya (from May 2015), all RTL and test bench files are Verilog 2001 or older and should be compatible with both ISE and Vivado (although compatibility has not been fully validated on ISE).*

Takeaways

Here are some of the key takeaways from this chapter:

- SDR is a complex radio communication system where components that have typically been implemented in hardware (e.g., mixers, filters, amplifiers, modulators/demodulators, detectors, etc.) are implemented through software, typically in an embedded system (like an FPGA) or PC.

- The SDR receiver tutorial that is provided by Red Pitaya is a great introduction to SDR technology.

- You need more than just an FPGA to pull off an SDR design. You also need an RF frontend and an ADC, which are electronic components outside of the FPGA device.

- The SDR# and HDSDR PC applications are cool ways to explore radio technology using your Red Pitaya SDR hardware.

FPGA Boards

Table A-1 lists some low-cost (under $200) development boards that use Xilinx FPGAs.

Table A-1 *Xilinx FPGA selection list*

Board	Cost	FPGA	GPIO	Interfaces	Free tools	Features
Papilio DUO (*http://bit.ly/ 1TyJF1j*) (Gadget Factory)	$87.99	Spartan-6 LX9	54-pin Arduino Mega Papilio Wings 1 Pmod	USB	Xilinx: ISE WebPACK Papilio: DesignLab	Onboard Arduino Micro
Papilio One 250K (*http://bit.ly/ 1TyJFhJ*) (Gadget Factory)	$37.99	XC3S250E	48-pin Papilio Wings	USB	Xilinx: ISE WebPACK Papilio: DesignLab	4 Mb SPI flash memory
Papilio One 500K (*http://bit.ly/ 1TyJIdp*) (Gadget Factory)	$64.99	Xilinx XC3S500E	48-pin Papilio Wings	USB	Xilinx: ISE WebPACK Papilio: DesignLab	4 Mb SPI flash memory

Board	Cost	FPGA	GPIO	Interfaces	Free tools	Features
Pipistrello LX45 H (http://bit.ly/1VE4ibt) (Saanlima Electronics)	$154.95	Spartan-6 LX45	48-pin Papilio Wings 1 Pmod	USB, DVI/HDMI, audio, SD card	Xilinx: ISE WebPACK	128 Mb SPI flash memory DRAM: 64 MB 7 LEDs
Mojo V3 (http://bit.ly/1VE4lnD) (Embedded Micro)	$74.99	Spartan-6 LX9	84-pin	USB	Xilinx: ISE WebPACK Arduino IDE	On board micro and flash 8 LEDs 1 reset button
Mimas (http://bit.ly/1VE4lUS) (Numato Lab)	$39.95	Spartan-6 LX9	70-pin	USB	Xilinx: ISE WebPACK	16 Mb SPI flash memory 8 LEDs and 4 switches
Saturn (http://bit.ly/1VE4oQu) (Numato Lab)	$119.95	Spartan-6 LX45	Up to 158-pin	USB	Xilinx: ISE WebPACK	DDR: 512 Mb 166 MHz LPDDR 128 Mb SPI flash memory
Waxwing (http://bit.ly/1SguMSF) (Numato Lab)	$199.95	Spartan-6 LX45	24-pin	USB, HDMI, VGA, 100BASE-T, SD card, JTAG, audio	Xilinx: ISE WebPACK	512 Mb LPDDR 128 Mb SPI flash memory 16 × 2-character LCD display 7-segment display 7 push buttons
FPGA Module 2.00b	$81.40	Spartan-6 XC6SLX16	100-pin	JTAG	Xilinx: ISE WebPACK	16 MB onboard flash memory

Board	Cost	FPGA	GPIO	Interfaces	Free tools	Features
(*http://bit.ly/ 1VE4pE6*) (ZTEX)						
USB-FPGA Module 2.01b (*http://bit.ly/ 1VE4nMy*) (ZTEX)	$108.90	Spartan-6 XC6SLX16	100-pin	USB, JTAG	Xilinx: ISE WebPACK	Cypress CY7C68013A EZ-USB FX2 microcontroller 128 Mb onboard flash memory 128 Kb EEPROM memory 2 Kb MAC-EEPROM 2 female 2 × 32 headers
USB-FPGA Module 2.04b (*http://bit.ly/ 1VE4s2D*) (ZTEX)	$130.90	Spartan-6 XC6SLX16	100-pin	USB, JTAG	Xilinx: ISE WebPACK	Cypress CY7C68013A EZ-USB FX2 microcontroller 64 MB DDR SDRAM 128 Mb onboard flash memory 128 Kb EEPROM memory 2 Kb MAC-EEPROM 2 female 2 × 32 headers

Board	Cost	FPGA	GPIO	Interfaces	Free tools	Features
Basys 2 (*http://bit.ly/ 1VE4uYl*) (Digilent)	$149.00	Spartan-3E	24-pin	USB, JTAG	Xilinx: ISE WebPACK	8 LEDs 4-digit 7-segment display 4 push buttons 8 slide switches PS/2 port 8-bit VGA port
Cmod S6 (*http://bit.ly/ 1VE4veY*) (Digilent)	$69.00	Spartan-6 XC6SLX4	48-pin	USB	Xilinx: ISE WebPACK	48-pin DIP form factor board 4 user LEDs 2 user buttons

Papilio AVR Loading

If you want to check out the operation of the Papilio AVR (Arduino microcontroller), there are a few additional steps you will need to complete.

Step 1: Power Up

First, plug your other USB-Micro cable into the Papilio module and move SW1 to the "up" position. You still need to have the FPGA USB cable connected to the USB-Mini connector. You don't have to change the PWRSEL jumper for this exercise.

You'll notice now that there are two Papilio DUOs listed in DesignLab's Tools→Port sub-menu: one for the FPGA and one for the AVR. You'll need to select the FPGA COM port first, as shown in Figure B-1.

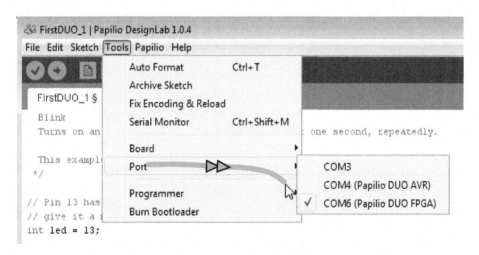

Figure B-1 *Select FPGA port*

Then you will need to select the AVR board type, as shown in Figure B-2. Select "AVR-USB" for this exercise. You do not want to select "AVR-No USB - ISP"!

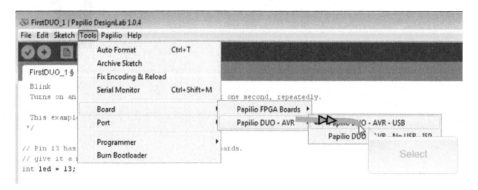

Figure B-2 *Select AVR board type*

 Need to Know

DesignLab lets you upload sketches to the AVR chip using the FPGA as an ISP programmer. The benefit of this is that it lets you use all of the code space available on the AVR chip. The Arduino bootloader eats up a couple of kilobytes of code space, and some sketches for the Arduino need that extra space. Some of the demos will not fit unless you use this method. The downside is that this method will completely wipe out the Arduino boot-loader, which means you will no longer be able to load sketches over the USB port. You can replace the bootloader, but it is a bit of a hassle, so Gadget Factory doesn't recommend using this method unless you run into a situation that requires it.

Step 2: Change Circuit Directive

Next, you need to change the circuit directive to remove the ZPUino soft core—but remember that you always need to have a circuit (FPGA bit file) associated with your sketch. You do this by simply using the "blank" circuit:

```
#define circuit blank
```

Figure B-3 shows what the sketch looks like with this directive added.

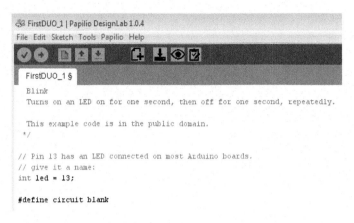

```
Blink
Turns on an LED on for one second, then off for one second, repeatedly.

This example code is in the public domain.
*/

// Pin 13 has an LED connected on most Arduino boards.
// give it a name:
int led = 13;

#define circuit blank
```

Figure B-3 *Blank circuit directive*

Step 3: Load FPGA Bit File

Now load the FPGA bit file to the module by clicking the Load Circuit icon in the Design-Lab toolbar as before. Wait for the "Done burning bit file" message to appear.

FYI

You still need a very basic FPGA image to route the reset signal through the FPGA to the AVR.

Step 4: Change COM Port

Change the COM port to the AVR port, as shown in Figure B-4.

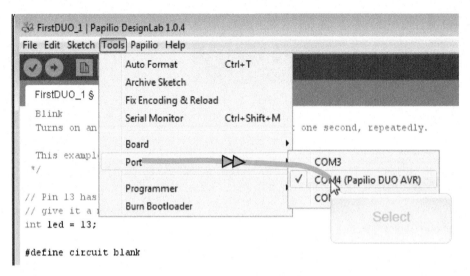

Figure B-4 *AVR port selection*

Step 5: Compile and Upload Sketch

Compile and upload the sketch (C program) to the ZPUino (Arduino soft core) so it can begin executing. You do this by clicking the Upload icon in the DesignLab toolbar. Wait for the "Done uploading" message and you should see the green LED blinking on your board again—only this time the code is executing in the AVR (Arduino microcontroller).

Serial Monitor

I like to add the serial monitor into my initial testing when using Arduino-compatible boards. It's a convenient way to check if the code you *think* is running is actually running. You can also use it for some simple code debugging techniques.

To do this, you will need to add a couple more lines of C code to your sketch. First, you'll need to configure your serial monitor to 9,600 baud in the setup section of your code:

```
Serial.begin(9600);
```

Then you need to add a print line statement inside your main loop. Make sure you have a semicolon at the end of each of these statements—that's a C syntax rule. You can write any text inside the quotes and it will be displayed in the serial monitor window of the IDE:

```
Serial.println("First Papilio DUO AVR Test");
```

Figure B-5 shows what the sketch should look like.

```
FirstDUO_1 | Papilio DesignLab 1.0.4
File Edit Sketch Tools Papilio Help

FirstDUO_1 §

  Turns on an LED on for one second, then off for one second, repeatedly.

  This example code is in the public domain.
 */

// Pin 13 has an LED connected on most Arduino boards.
// give it a name:
int led = 13;

#define circuit blank

// the setup routine runs once when you press reset:
void setup() {
  // initialize the digital pin as an output.
  pinMode(led, OUTPUT);
  Serial.begin(9600);
}

// the loop routine runs over and over again forever:
void loop() {
  digitalWrite(led, HIGH);   // turn the LED on (HIGH is the voltage level)
  delay(1000);               // wait for a second
  digitalWrite(led, LOW);    // turn the LED off by making the voltage LOW
  delay(1000);               // wait for a second
  Serial.println("First Papilio DUO AVR Test");
}
```

Figure B-5 *Adding serial monitor code*

Compile and upload your sketch as before, then open the serial monitor from the Tools menu of the IDE (Figure B-6).

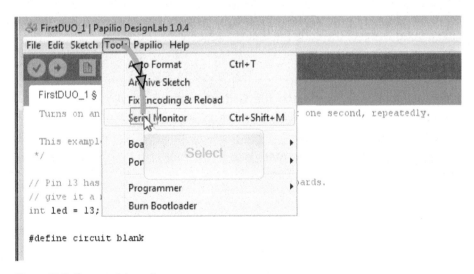

Figure B-6 *Open serial monitor*

You should see the text you coded in the serial monitor window (Figure B-7).

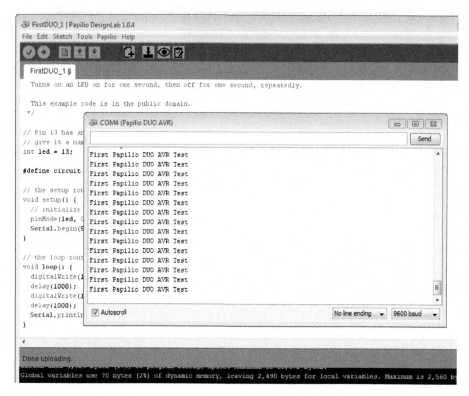

Figure B-7 *Serial monitor window*

You can use this same code in the ZPUino. I typically change the text inside the quotes, which gives me a good visual of the code file that I'm actually executing.

Additional Resources

I highly recommend that everyone watch the Gadget Factory's Papilio DUO Overview video and DesignLab Tour video (*http://bit.ly/1KK2zkx*).

You should also review the DUO QuickStart Guide (*http://bit.ly/1KK2tcC*) and watch the FPGA video (*http://bit.ly/1KK2tcC*) and the AVR video (*http://bit.ly/1KK2xsY*), all of which are available on the Gadget Factory website.

Text and Code Editor

It is a good idea to have a text and code editor in your FPGA toolbox. There will be times when you will need to look at an HDL code file outside of the Xilinx ISE WebPACK. For those of you who will be doing most of your FPGA exploration on a Windows PC, you will need an editor that is more sophisticated than Notepad. I like to use the freeware editor PSPad (*http://www.pspad.com/index_en.html*) when I'm doing any type of code development on a Windows platform; it has a fairly rich feature set.

There are many good text and code editors out there. Wikipedia (*http://en.wikipedia.org/wiki/List_of_text_editors*) has a good list of many of them.

When I'm working on a Linux platform I use gedit (*https://wiki.gnome.org/Apps/Gedit*), another freeware program. This is a basic text/code editor with a reasonably good graphical user interface. Now, for many of the hardcore hardware and software developers using the Linux OS platform, the whole topic of code editors is hallowed ground. I have been in some development labs where just the mention of gedit is like speaking blasphemy to the gods of the command line. Seriously, I have gotten into the crossfire of many heated debates between the opposing forces of the *vi* and Emacs cults. The bottom line is, choose whatever editor floats your boat.

Index

About the Author

David Romano recently founded Tri-Tech Pathways Inc. to bring STEM education to students with a real-world industry perspective. He is a proven technical leader whose engineering career has spanned over 25 years and multiple high-tech companies, including Raytheon, Motorola, HP, Intel, and two start-up companies. He is the coauthor of multiple technology patents and is currently the president and CEO of Tri-Tech Pathways Inc.

David is currently pursuing a doctorate degree in education and he holds degrees in electrical engineering and theology. He is actively involved in STEM advisory and teaching roles. He is also a member of the International Society for Technology in Education (ISTE) and the Computer Science Teachers Association (CSTA).

David is an avid motorcyclist who enjoys both on-road and off-road riding; he loves hiking and snowshoeing in the mountains of New England and is also a certified scuba diver.

Colophon

The cover image was created by Shawn Wallace. The cover fonts are Benton Sans and Soho Pro. The text font is Adobe Myriad Pro; the heading font is Benton Sans; and the code font is Dalton Maag's Ubuntu Mono.

CPSIA information can be obtained
at www.ICGtesting.com
Printed in the USA
BVOW10s1405150516
448158BV00003B/15/P